Applied Ontology Engineering in Cloud Services, Networks and Management Systems

J. Martín Serrano Orozco

Applied Ontology Engineering in Cloud Services, Networks and Management Systems

J. Martín Serrano Orozco
National University of Ireland Galway - NUIG
Digital Enterprise Research Institute - DERI
Lower Dangan - IDA Business Park
Co. Galway, Galway City, GALCT, Ireland
martin.serrano@deri.org

ISBN 978-1-4899-8587-3 ISBN 978-1-4614-2236-5 (eBook)
DOI 10.1007/978-1-4614-2236-5
Springer New York Dordrecht Heidelberg London

Softcover reprint of the hardcover 1st edition 2012

Printed on acid-free paper

Springer is part of Springer Science+Business Media (www.springer.com)

To Jaime and Oliva…

To Irene, Fernando and Luis…

To Karla, Diego and Omar…

Preface

In the Information and Communications Technology (ICT) sector where metadata standards are proliferating at unprecedented levels and automated information management systems need to collect and process information from, every day, sensors, devices, applications, systems, etc., the interoperability of the information and the knowledge exchange is identified as one of the major challenges. Furthermore, this necessity continues increasing as result of the everyday new demands for smarter services and the large amount of information captured and generated electronically, and also for today's trend about the information being processed and stored in cloud applications.

The introduction of these cited new demands bring as clear consequence the inevitable diversification of information, then as a predictable requirement, efficient data collection and optimal processing mechanism have to be designed and implemented and service providers, operators and infrastructure managers have to operate with many different standards and vocabularies. The use of semantic techniques to face up this requirement sounds as an optimum approach providing advantages in terms of facilitating data and information mapping. However those semantic techniques are becoming very costly, in terms of computing resources and time-response factors, resulting in loss of business opportunities because existing systems are not adaptable enough to those scalability demands.

If this cross-domain and information sharing problem is translated into the communications domain, as a result, it is necessary to deal with the most common scenario where computer network infrastructures collect data with the objective of supporting diverse communication services and applications. In communications systems, the problems about interoperability of data and information exchange are evident, and even more in recent days when the IT sector and its demands about convergence within heterogeneous communication systems, these problems are increased exponentially, emphasizing the need for proper semantic techniques and methodologies to facilitate information exchange processes.

ICT market is evolving towards more interconnected platforms and demanding integrated service solutions; for example, cloud computing emerges as one of the most promising and at the same time more adaptive computing services, generating the multiplicity of middleware approaches and infrastructure development.

Cloud computing concentrates on tackling not only the complex problems associated with service cost reduction demands but also those related to network performance, efficiency and technological flexibility [Greenberg09]. It is true there is a market boom generated about cloud computing solutions; it rises exponentially when the economic atmosphere drives the world economies to a more service revenue with less investment in technology. However, beyond the service advantages and marketing revenue cloud-based solutions can provide, it is absolutely necessary to have well-identified service challenges, technology requirements and clear customer and service provider necessities to specifically and correctly solve root problems.

As a main common requirement when diverse systems are interacting, the linked data and exchange of information are considered as crucial features. In this diverse and complex set of service and technological requirements, about information exchange and systems integration respectively, the role of management systems and other next generation applications and networking services supporting the mentioned integration is most likely catalogued as a set of alternatives.

A traditional scenario example to understand this complexity can be studied in autonomic systems management [see work from the Autonomic Communications Forum (ACF)]. Acting as a root problem, autonomic networks are not able to handle the broad diversity of information from resources, devices, networks, systems and applications. Particular interest focuses on exchanging information between different stake holders (autonomic components or autonomic layers) when it is necessary; however as described, there is no capability to exchange pieces of such information between the different systems participating in the autonomic solution.

The convergence of software and networking solutions can provide solutions for some of the complex management problems present in current and future Information and Communications Technologies (ICTs) systems. Current ICT research is focused on the integrated management of resources, networks, systems and services. This can be generalized as providing seamless mobility to, for example, personalize services automatically. This type of scenarios requires increased interoperability in service management operations.

Integrated management and cross-layer interactions involve both the transmission capabilities from network devices and the context-aware management services of the middleware environment. Transmission capabilities influence the performance of the network, while middleware impacts the design of interfaces for achieving data and command interoperability.

Integrated management refers to the systematization of operations and control of services in networks and systems. Cross-layer refers to the joint operation of the physical, link, management and service layers, and context-awareness refers to the properties that make a system aware of its user's state, the goals of the user and operator, and the state of the network environment. This awareness helps the systems to adapt their behaviour according to applicable business rules, all the while offering interoperable and scalable personalized services. To do this, different data models are required in NGN and Internet solutions, due to the inherent heterogeneity of vendor devices and variability in the functionality that each device offers.

Typical solutions have attempted to provide middleware to mediate between a (small) subset of vendor-based solutions, while research has investigated the use of a single information model that can harmonize the information present in each of these different management data models. Industry has not yet embraced the approach, since this research typically does not map vendor-specific functionality to a common information model.

To alleviate this problem in ICT systems, the convergence of software solutions and managing systems controlling the networking infrastructures must provide alternate solutions in short incubation periods and with high scalability demands. Additionally and as result on the increasing demand to implement cloud solutions, it is not difficult to understand why ICT research is focusing on the integrated management of resources, networks, systems and cloud solutions or cloud services and on the way to exchange data standards facilitating this information interoperability and systems integration labor. These challenges in terms of realistic services and applications can be generalized as looking forward for providing seamless mobility services to, for example, personalize services automatically.

This book focuses on Ontology Engineering and its applications in service and network management systems. This book aims to act as a reference book defining application design principles and methodological modeling procedures to create alternative solutions to the scientific and technological challenge of enabling information interoperability in cross-domain applications and systems, examples in managing cloud services and computer network systems are included.

In today's ICT systems, the enormous amount of information and the increasing demands for optimally managing them generates the necessity of rethinking if current information management systems can cope to these demands and what are the best practices to make more efficient the telecommunications services in computer networks, particularly in times where everything is migrating to cloud-based systems. Thus this book is expressly an invitation to explore and understand the basic concepts, the applications and the consequences and results about applying ontology engineering in cloud services, networks and management systems

J. Martín Serrano Orozco

Who Should Read This Book?

Today engineering and computer science professionals (infrastructure architects, software developers, service designers, infrastructure operators, engineers, etc.) are facing up as never before the challenge of convergence in software solutions and technology; a clear example of this trend is the integrated management in the Information and Communications Technologies (ICT) domain.

As a scientific concept in ICT, interoperability of the information sounds like a feasible solution for one of the traditional challenges in this area, and the advantages of having implemented are positives, however it is still far the day this problem can be solved by simple software implementation activity and relatively ad hoc technological deployment, it is mostly act itself as a limitation due this convergence bring a set of requirements acting as restrictions when heterogeneous technologies are in place.

However software engineering and information management professionals are transforming their vision of different disciplines to generate a unified converged and multidisciplinary engineering perspective. Today's professionals in ICTs need to create efficient tools that range from network infrastructure representing data up to information modeling formalisms representing knowledge. These professionals need assistance and guidance to pursue a better understanding about this convergence, and even more, the emerging tools and mechanism must be applied to create inter-domain applications, as result of this continuous and multidisciplinary process.

This book focuses on the necessity to understand clearly the impact of the information in communication systems, the features and requirements for software applications design and the trends in large scale information management systems based in cloud, this book describes and exemplifies, rooted on scientific research and point towards implemented solutions, the processes of modeling and managing information. A guide in the form of a cookbook for enabling services using ontology-based information and data models that can be exchanged between different information management levels and applications. It make references to methodological approaches rooted in the ICT area, creating solution(s) for information interoperability problems between network and service management domains in the era of cloud computing.

This book is aimed for the wide ICT-sector people, engineers in general, software developers, students, technology architects and people with knowledge on semantic principles, semantic Web formal languages and people with knowledge and rooted in Internet science and telecommunications.

This book is addressed to those who realistically see the interaction of network infrastructure and software platforms as a unified environment where services and applications have a synergy exchanging information to/for offering cognitive applications commonly called smartness or intelligence in computing and cognitive or awareness in telecommunications.

This book is suitable to be read and/or studied by students with strong basis in communications, software engineering or computer science or any other related disciplines (A level of engineering studies is required or its equivalent in different knowledge areas). This book is not intended to be a text book, but if well studied can provide engineering methodologies and good software practices that can help and guide the students to understand principles in information and data modeling, integrated management and cloud services.

This book is a scientific tool for those active professionals interested in the emerging technology solutions focused on Internet science and semantic Web into the communications domain, a very difficult combination to find in current literature references, where a high degree of focus and specialization is required. In this book, as a difference from other literature references, the underlying idea is to focus on enabling inter-domain and intra-domain interactions by augmenting information and data models with semantic descriptions (ontological engineering). There are realistic scenarios where the techniques described in this book have been applied.

Finally, this book is not aimed at students of different disciplines beyond those considered in the framework of IT and Communications (ICT's), but it is suitable for those students and people with general interest in service applications, communications management, future Internet and cloud computing principles.

What is Covered in this Book?

This book concentrates on describing and explaining clearly the role ontology engineering can play to provide solutions tackling the problem of information interoperability and linked data. Thus, this book expressly introduces basic concepts about ontology engineering and discusses methodological approaches to formal representation of data and information models, facilitating information interoperability between heterogeneous, complex and distributed communication systems. In other terms, this book will guide you to understand the advantages of using ontology engineering in telecommunications systems.

This book discusses the fundamentals of today's ICT market necessity about convergence between software solutions and computer network infrastructures. This book introduces basic concepts and illustrates the way to understand how to enable interoperability of the information using a methodological approach to formalize and represent data by using information models. This book offers guidance and good practices when ontology engineering is applied in cloud services, computer networks and management systems.

Acknowledgements

To my family for their incomparable affection, jollity and always encouraging me to be creative, and for their enormous patience during the time away from them, invested in my personal formation, and for their understanding about my professional life and its consequences, everything I am, is the result of your teachings.

To all my friends for their comprehension when I had no time to spend with them and for the attention and the interest they have been showing all this time to keep alive our friendship; the sacrifice has been well rewarded.

To all my colleagues, students and friends in WIT-TSSG (Waterford, Ireland) for patiently listening with apparent attention the descriptions of my work and for the great experiences and the great time and working place they provided to culminate this personal project in my professional life. In particular, thanks for the support from all people who believed this book would be finished and also to those who didn't trust on it, because thanks to them I was more motivated to culminate the project. Sincerely to you all, thanks a million.

To my friends and colleagues Joan Serrat Fernandez and John C. Strassner for their support, for the time they invested in my professional formation and to believe in my outlandish ideas and mainly for their unconditional help and guidance all the time.

J. Martín Serrano Orozco

Acknowledgements

To my colleagues and friends in NUIG-DERI, Stefan Decker and Manfred Hauswirth for their guidance and the great support on my research activity; Josiane Xavier Parreira, Christian von der Weth, Danh Le Phuoc, Mario Arias Gallego, Anh Le Tuan for their efforts and support all the time; All staff and students for their enthusiasm spent on job's activities which contributes to create the great collaborative-friendly atmosphere in DERI (Galway, Ireland).

To my collegues and friends in EPFL, Karl Aberer, Thanasis G. Papaioannou, Sofiane Sarni (Lausenne, Switzerland); AIT, John Soldatos, Aristodemos Pnevmatiakis, Nikos Kefalakis (Athens, Greece); Fraunhofer-Gesellschaft, Gerhard Sutschet, Reinhard Herzog (Karlsruhe, Germany); SENSAP, S.A., Panos Dimitropoulos, Nikos Zarokostas, Achilleas Anagnostopoulos (Athens, Grece); UKITA IT Asssociation, Charles Huthwaite, Tom Brookes (Birmingham, United Kingdom); CSIRO, Dimitros Georgakopoulos, Arkady Zaslavsky, Ali Salehi, Armin Haller (Australia) for their professionalism, share their knowledge and the great support all the time to make the project a successful experience.

To my colleagues and friends in NUI-Maynooth, Afra Kerr, Stefano De Paoli and NUI-Maynooth Hamilton Institute, Douglas Leith, Laetitia Chapel, David Malone (Maynooth, Ireland); NUI-Limerick, Mikael Fernström, Cristiano Storni (Limerick, Ireland); WIT-TSSG, Michaél Ó Fóghlú, Dmitri Botovich, Sasitaran Balasubramaniam, Chamil Kulatunga, Tom Pfifer (Waterford, Ireland) for their patience and open mind to discuss ideas and for sharing their experiences.

To my colleagues and friends in UCD, Liam Murphy, John Murphy, Philip Perry, Viliam Holub, Trevor Parsons (Dublin, Ireland); TCD, Declan O' Sullivan, John Keeney, Owen Conlan, Rob Brennan, Kevin Feeney (Dublin, Ireland); UCC, Simon Foley, William Fitzgerald (Cork, Ireland); WIT-TSSG, Willie Donnelly, Sven Van der Meer, Brendan Jennings, Martin Johnsson, Ruari de Frein, Lei Shi for their guidance and work advice and their support all the time.

To my colleagues and friends in Telefonica I+D, Juan Manuel Sánchez, José Fabian Roa Buendía, Jose Antonio Lozano (Madrid, Spain); Takis Papadakis, Vodafone (Athens, Greece); Panos Georgatsos, Takis Damilatis, Dimitrios

Giannakopoulos, Algonet, S.A. (Athens, Greece); VTT Arto Tapani Juhola, Kimmo Ahola, Titta Ahola (Espoo, Finland); Technion, Danny Raz, Ramic Cohen (Haifa, Israel); UCL, Cris Todd, Alex Galis, Kun Yang, Kerry Jean, Nikolaos Vardalachos (London, UK); NTUA, Stavros Vrontis, Stavros Xynogalas, Irene, Sygkouna Maria Schanchara and UPC, Joan Serrat Fernandez, Javier Justo Castaño and Ricardo Marin Vinuesa (Barcelona, Spain) for their direct and indirect contributions when we were together discussing ideas and for sharing their always valuable point of view.

J. Martín Serrano Orozco

This is a contribution to the EU Project EU-IST FP7-ICT-2011-7-287305 – OpenIoT
Open Source blueprint for large scale self-organizing cloud environments for IoT
EU CORDIS – Community Research and Development Information Service
EU IST – Information Society Technologies,
European Comission, Europe

This is a contribution to the Digital Enterprise Research Institute – CSET DERI
Centre for Science, Engineering and Technology SFI-08-CE-I1380
National University of Ireland Galway, NUIG,
Science Foundation Ireland,
Galway, Ireland

This is a contribution to the SFI-FAME SRC – FAME SRC
Strategic Research Cluster SFI-08-SRC-I1403
SFI – Science Foundation Ireland,
Waterford, Ireland

This is a contribution to the HEA-FutureComm Project – FutureComm
High Education Authority Ireland
Waterford Institute of Technology
Waterford, Ireland

J. Martín Serrano Orozco

Contents

Abbreviations

3GPP	3rd generation partnership program
ACF	Autonomic communications forum
ADB	Associated data base
ADL	Application definition language
AI	Artificial intelligence
AIN	Advanced intelligent networks
AN	Active networks
ANDROID	Active network distributed open infrastructure development
ANEP	Active network encapsulation protocol
ANSI	American National Definition Standards Institute
API	Application programming interface
ASP	Application service provider
BSS	Business support systems
CAS	Context-aware service
CCPP	Composite capabilities/preference profiles
CD	Code distributor
CDI	Context distribution interworking
CEC	Code execution controller
CIDS	Context information data system
CIM	Common information model
CLI	Command line interface
CMIP	Common management information protocol
CMIS	Common management information service
COPS	Common open policy service
CORBA	Common object request broker architecture
CPU	Central processing unit
DAML	DARPA Agent Markup Language
DAML-L	DAML logic
DAML+OIL	DAML-L + OIL language fusion

DCOM	Distributed component object model
DB	Database
DiffServ	Differentiated services
DEN	Directory enabled networks
DEN-ng	Directory enabled networks-next generation
DMC	Decision-making component
DMI	Desktop management interface
DMTF	Distributed management task force
DNS	Domain name service
DNSS	Directory naming and specification service
DTD	Document type definition
EAV	Entity-attribute value
EE	Execution environment
eTOM	Enhanced telecommunication operations map
EU	European Union
FPX	Framework program—X the number referring the program
GDMO	Guidelines for the definition of managed object
GIS	Geographic information systems
GPRS	General packet radio system
GSM	Global system for mobile communications
GUI	Graphical user interface
HTML	Hyper text markup language
HTTP	Hyper text transfer protocol
IDL	Interface definition language
IEC	International engineering consortium
IEEE	Institute of electrical and electronic engineers
IETF	Internet engineering task force
IFIP	Internet federation for information processing
IM	Information model
IMO	Information model object
ISL	Invocation service listener
IN	Intelligent networks
IntServ	Integrated services
IP	Internet protocol
IRTF	Internet research task force
ISO	International Standardization Organization
ISP	Internet service provider
IST	Information society technologies
IT	Information technologies
ITC	Information technologies and communications
ITU-X	International telecommunication unit-section
JavaSE	Java platform standard edition
JavaEE	Java platform enterprise edition

JavaME	Java platform micro edition
JIDM	Joint inter-domain management
JMF	Java media framework
JVM	Java virtual machine
KIF	Knowledge interchange format
KQML	Knowledge query and manipulation language
LBS	Location-based services
LAN	Local area network
LDAP	Lightweight directory access protocol
MAC	Medium access control
MDA	Model-driven architecture
MIB	Management information base
MPLS	MultiProtol label switching
MOF	Managed object format
NGI	Next generation Internet
NGN	Next generation networks
NGOSS	Next generation operation system support
OCL	Object constraint language
ODL	Object definition language
OIL	Ontology interchange language
OKBC	Open knowledge base connectivity
OMG	Object management group
OS	Operating system
OSA	Open service access
OSI	Open systems interconnection
OSM	Ontology for support and management
OSS	Operations support systems
OWL	Web ontology language
P2P	Peer-to-pair
PAL	Protegé axiom language
PBM	Policy-based management
PBNM	Policy-based network management
PBSM	Policy-based service management
PC	Personal computer
PCC	Policy conflict check
PCIM	Policy core information model
PCM	Policy consumer manager
PC	Policy consumer
PDA	Personal digital assistant
PDP	Policy decision point
PE	Policy editor
PEP	Policy enforcement point
PPC	Policy conflict check

PGES	Policy group execution strategy
PM	Policy manager
PPIM	Parlay policy information management
PSTN	Public switched telephone network
QoS	Quality of service
RDF	Resource description framework
RDFS	RDF schema
RFC	Request for comment
RMI	Remote method invocation in JAVA technologies
RSVP	Resource reservation protocol
RM-ODP	Reference model for open distributed processing
RM-OSI	Reference model for open system interconnection
SAM	Service assurance module
SDL	Specification and description language
SDK	Software development kit
SGML	Standard generalized markup language
SID	Shared information model
SLA	Service level agreement
SLS	Service level specifications
SLO	Service logic
SMTP	Simple mail transfer protocol
SNMP	Simple network management protocol
SOA	Service-oriented architecture
SOAP	Simple object access protocol
SSL	Secure socket layer
SP	Service provider
TCP	Transmission control protocol
TEManager	Traffic engineering manager
TMF	Telecommunications management forum
TM Forum	TeleManagement forum
TMN	Telecommunications management network
TOM	Telecommunications operations map
UDDI	Universal description, discovery and integration
UI	User interface
UML	Unified modelling language
UMTS	Universal mobile telecommunication system
URI	Uniform resource identifier
URL	Uniform resource locator
VAN	Virtual active network
VE	Virtual environment
VoIP	Voice over IP
VM	Virtual machine
VPN	Virtual private network

W3C	World wide web consortium
WBEM	Web-based enterprise management
WAN	Wide area network
WWW	World wide web
XDD	eXchange data definition
XMI	XML metadata interchange
XML	eXtensible markup language
XSD	XML schema data-types
XSL	eXtensible stylesheet language
XSLT	eXtensible stylesheet language transformation

Y

Z

List of Figures

List of Tables

Chapter 1
Convergence in IT and Telecommunications

1.1 Introduction

This chapter describes general trends in telecommunications and makes references to semantic Web principles used in service and network management systems. In today's communication systems, there is a trend of using the semantic Web to simplify operations and execute either service and network control operations using pieces of monitored information and data. The operations are described and represented by pieces of information and particular data models to then execute networking and service management operations. Semantic drawbacks in IT systems are inherent in those operations, due to the network service and end user disassociation in information. However, such drawbacks must be addressed in order to satisfy business goals pursuing the integrated management of the services in next-generation networks (NGNs).

In other words, this chapter presents issues about convergence in IT and telecommunications, reviews information interoperability problems when using context information and looks at the alternatives to achieve integration of data with semantic-based models and linked data mechanisms. From this perspective, it is proposed that ontology engineering techniques can be adopted as formal mechanisms in order to solve some of the management and interoperability problems caused by the use of diverse information and data models. So, when integrating context information with service requirements and management data, solutions in the form of middleware have to be designed and implemented.

Recently, emphasis on virtual infrastructures has risen, as a result of its versatility to execute multiple tasks and their relatively easy configuration and deployment. This approach results in multiple advantages in semantic content and pervasive applications, particularly when communications services are being managed, for example a service can automatically be configured if it can access the profile descriptions of a new user to offer personalized services and the infrastructure supporting such services can be adapted and its operation modified.

The convergence of software and networking solutions can provide solutions for some of the complex management problems present in current and future information

J.M. Serrano Orozco, *Applied Ontology Engineering in Cloud Services, Networks and Management Systems*, DOI 10.1007/978-1-4614-2236-5_1,

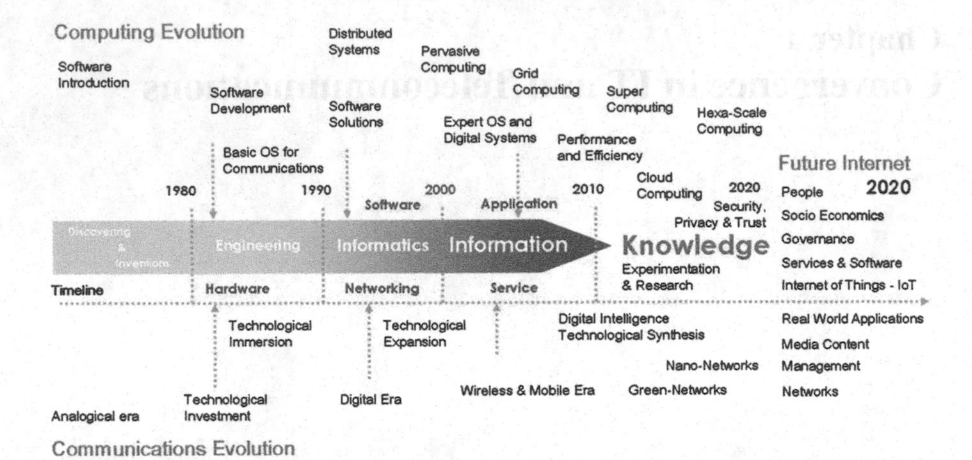

Fig. 1.1 Convergence in software and communications towards an ICT integration model

technologies and communications (ITC) systems. Current ITC research is focused on the integrated management of resources, networks, systems and services. This can be generalized as providing seamless mobility to, for example personalize services automatically. This type of scenarios requires increased interoperability in service management operations. In addition, by using cloud-based computational resources (mainly virtualization) and ontology engineering for supporting pervasive services and management tools (policy-based management) for the emerging cloud computing area, integrated solutions can be promptly tested. Figure 1.1 represents this concept where convergence of software and technology is depicted.

In addition, related with convergence of interdisciplinary areas, this section focuses on describing and studying the convergence in IT and telecommunications with an emphasis on semantic Web and services management for managing Internet services by using pieces of information (the so-called context awareness) and end-user requirements by using software-oriented architectures (SOA) and policy-based management paradigms (service and network management). Likewise, this chapter introduces in a general way the emerging area of cloud computing and provides an introductory discussion about it. To date, the technology advances in communication services have evolved following the two main paths. The first is driven by the need of common devices with embedded technology and connectivity to become progressively smaller and more powerful; the second is the cooperation of diverse systems for supporting multiple services, due to the convergence and interaction of pervasive computing and communications systems. As a result of this evolution, it is essential for information to be as interoperable as possible in current and future applications and services. In particular, the exchange and reuse of information is mandatory for managing service operations.

Particularly, this chapter concentrates in analyzing and studying combined solutions to facilitate information interoperability among heterogeneous, complex and distributed communication systems when managing pervasive services.

Finally, rooted in the ITC, rather to define terminology and propose semantic interoperability problems as an alternative for network and service management in cloud systems, this chapter discusses about service-oriented architectures, and the role management information contained as services description and described as policy information (data models) can be used for providing extensible, reusable, common manageable knowledge layer for better management operations in cloud computing.

The organization of the rest of this chapter is as follows. Section 1.2 describes the trends in the management of communication services and semantic Web areas. It describes the convergence between management and middleware, explaining the need for better services management systems, and introduces ontology engineering as a means to semantically support service management in autonomic communications, which can be used to integrate cloud computing solutions to create and deploy new embedded services in virtual environments.

Section 1.3 introduces Internet design trends and how the future of the Internet is being tracked and re-shaped by emerging technologies in telecommunications. This section describes the stages about software services and telecommunications. The stages described in this section constitute a research activity and can be considered as parts contributing to the state of the art about the convergence in IT and telecommunications. The proposal of those stages are depicted and described; this provides a detailed understanding of the path taken from the start of research activity to the delivery of this book.

Section 1.4 introduces concepts related to SOA and basis of cloud computing. Interoperability is an inherent feature established in SOA for addressing heterogeneous, complex and distributed issues, where management operations are also required. Commonly in SOA design, the interoperability plays a protagonist role, however studied development common practices reflects that implemented approaches traditionally espouse a strict inter-functionality and cross-layered interactions has been left beside, however in this section, current management systems the broad diversity of resources, devices, services and systems of converged networks to be applicable to NGNs and pervasive services applications are considered.

Finally, Sect. 1.5 presents the conclusions in order to summarize about the fundamentals and trends on Semantic Web, Telecommunications Systems and Cloud Computing introduced and discussed in this chapter. The aim of this section is to establish a general understanding about the interoperability and interaction between the areas described in this chapter.

1.2 Semantic Web and Services Management

The process of integrating computing solutions to create and deploy new embedded services in pervasive environments results in the design and development of complex systems supporting large number of sensors, devices, systems and networks, where each of which can use multiple and heterogeneous technologies. In the area of Internet and semantic Web, Web sensors bring the concept of formal descriptions associated to features in the Web to generate intelligent content. The intelligent content is structured information usually represented by a formal language that makes the information easy accessible and able to be used by different applications either end user or infrastructure related. In the area of the communications management, the concept of seamless mobility associates scenarios where people configure their personalized services using displays, smart posters and other end-user interaction facilities, as well as their own personal devices as a result of information exchange.

The inherent necessity to increase the functionality of Web services in the Internet is particularly motivated by both the necessity to support the requirements of pervasive services and the necessity to satisfy the challenges of self-operations dictated by the communications systems [Kephart03]. Web services requirements are headed by the interoperability of data, voice and multimedia using the same (converged) network. This requirement defines a new challenge: the necessity to link data and integrate information. Specific scenarios become evident when management instructions are used for expressing the state of users and defining services performance, and such instructions are considered as pieces of information able to be exchanged between different management systems. Ideally, as a result of this interaction, i.e. semantic Web services, service management is possible dynamically to adapt the services and resources that they provide to meet the changing needs of users and/or in response to changing environmental conditions. This adaptation is essential, as each day, more complex services are required by consumers and the main driver of those services is the Web, which in turn requires more complex support systems that must harmonize multiple technologies in each network and semantic information from the Internet.

A more complete visionary approach about service management promise new, user-centric applications and services. NGNs and services require information and communications systems able to support information services and especially applications able to process pieces of information. Information plays the important role of enabling a management plane where data is used in multiple applications with no restrictions and it is able to be adaptive according to services and resources that it is designed to be offering. The processes of linking data and information management services, by using information data models, pursues the common objective of changing the demands of the user, as well as adapt the changing environmental conditions, in the form of interoperable information, thus helping to manage business, system and behavioural complexity [Strassner06a].

The information is dynamic; therefore, the efficient handling and distribution of data and information to support context-awareness is not a trivial problem, and has generated study dating back to the first time that a simple unit of information, known as context model, was proposed [Chen76]. The multiple advantages derived from modelling context have attracted much attention for developing context-aware applications, generating diverse approaches and turning ubiquitous computing into what is currently known as pervasive computing. Nevertheless, most research has focused on realizing application-specific services using such information. Such application-specific uses of information are rarely portable and usable by other applications or services, since they usually utilize proprietary representations and languages. This adversely impacts the management of the services and makes cross-layer interoperability harder to achieve.

In addition, the use of an information model enables the reuse and exchange of service management information. Some examples and initiatives of using common information model include the initiative CIM/WBEM (common information model/ web-based enterprise management), [DMTF-CIM] from the DMTF (Distributed Management Task Force, Inc.) and broadly supported by the shared information model (SID) of the TMF (TeleManagement Forum). However, neither has been completely successful, as evidenced by the lack of support for either of these approaches in network devices currently manufactured. This indicates that SID model lacks the extensibility to promote the interoperability and enhance its acceptance and expand its standardization. Other examples in the level of operation and applications with other technologies are Microsoft DCOM (distributed component object model—Microsoft) or RMI (remote method invocation—Sun Micro Systems) which are being used in many applications. Even those initiatives do not allow sharing the information with each other technologies freely.

The vision of the future, which enables societies to use computing systems to "transparently" create pervasive services automatically, requires associated information and networks management systems that are able to support such dynamic service creation. Under this ambitious panorama, the multiplicity and heterogeneity of technologies used, such as wireless, fixed networks and mobile devices that can use both, is a barrier to achieving seamless interoperability. This vision uses formal ontology languages to get a formal representation, describe content, share and reuse information. Thus, this initiative creates the basis to find the best way to integrate context information into service management operations using formal mechanisms; it is also used to propose extensions for the support of information sharing and seamless mobility scenarios.

1.3 Internet Services and Telecommunications

Traditionally, the management of communications systems (services, computing components and network resources) has been done by humans. Those management actions are today so complex that human operators need to be assisted by appropriate computing systems capable of deal with such complexity. With the broad development of Web technologies, the management area has also been influenced, and new proposals and technologies have contributed to enhance existing management systems supporting the idea of integrated management and cross-layer interactions as a result of the immersion of computing systems in high-scale communications systems management.

Integrated management refers to the systematization of operations and control of services in networks and systems. Cross-layer refers to the joint operation of the physical, link, management and service layers and information to the properties that make a system aware of its user's state and the state of the network environment. This awareness helps the system to adapt its behaviour (e.g. the services and resources that it offers at any one particular time) according to desired business rules, all the while offering interoperable and scalable personalized services.

Integrated management and cross-layer interactions involve both the transmission capabilities from network devices and the information management services of the middleware environment (software supporting network and service operations and tasks). Transmission capabilities influence the performance of the network. Hence, their impact on the design of new protocols, and the adaptation of existing protocols to suit these capabilities, need to be studied by modelling and/or simulation of systems. Middleware development impacts the design of interfaces for achieving the interoperability necessary in services.

Actually, there are many initiatives that are breaking with the models of how fixed and mobile communication networks operate and provide services for consumers. Those initiatives are founded on existing basis and many times acting as "standards" regarding NGN; however, multiplicity of "standards" is not a good envision for providing services. Examples of those initiatives create a background of this book and are available in [Brown96b], [Brown97] and [Chen00] as well as in industry applications [Brown96c], [Brumitt00] and [Kanter02], just to cite some examples. These efforts try to move from a world where the networks are designed and optimized around a specific technology, service and/or device towards a world in which the user is at the centre of his/her communications universe. In this new communications world, resources and services become network- and device-agnostic. NGN services are thus no longer device-, network-, and/or vendor-centric as current services are; they now are user-centric.

The new user-centric vision, where users define preferences and personalize services, brings itself inherent complex problems of scalability and management capacity. Acting as an particular requirement for new design approaches; it is required formal and extensible information model(s) be used according to specific service management requirements, in order to enable scalability in the systems. Internet services focus on the information management part to contribute to the support of these kinds of applications. Even standards, such as the SNMP (simple network management protocol) [IETF-RFC1157], and its updated versions [IETF-RFC2578] have failed in standardizing most of the key information required for management interoperability. This is because there is no fundamental interests for device vendors to follow a standard describing how their devices must work, likewise neither standard satisfy every vendor information requirement. Hence, Internet and their services will be harder to manage than current applications and services, since Internet applications and services are built from and are supported by more diverse networks and technologies. This causes control plane to be made up of different types of dissimilar control functions. Therefore, a management plane supported by complex management systems is needed to coordinate the different types of control planes, ensuring that each application, service, network and device play its role in delivering their functionality and all together constitutes the so-called end-to-end service.

1.4 SOA and Cloud Computing

Interoperability among heterogeneous, complex and distributed communication systems requires a new management approach. Traditionally, management systems offer solutions with strict layering of functionality, where from a design conception, cross-layered interactions are not considered or defined as crucial part of the application(s), which it blocks the open exchange of information between different applications. In today's management systems design the broad diversity of resources, devices, services, systems and the exchange of data and integration of information in heterogeneous networks is to be applicable. In NGN's and Internet services, all those new features are a must to be considered [TMN-M3010]. As an inherent characteristic, previous management systems do not support a large spectrum of devices, such as wearable computers and specialized sensors. Furthermore, previous systems are not provided with embedded technology/connectivity, which is used to make new types of networks that provide their own services (e.g. simple services supporting other, more complex, services).

These scenarios aim to provide more pervasive services and require linked data and sharing features to enable information exchange [Sedaghat11]. Such scenarios are typified by a broad mixture of technologies and devices that generate an extensive amount of different types of information, many of which need to be shared and reused among the different management components with different mechanisms of each network. This requires the use of different data models, due to both the nature of information being managed and the physical and logical requirements of application-specific systems and beyond that with security levels to guarantee service reliability [Waller11]. However, information/data models do not have everything necessary to build up this single common interoperable sharing of knowledge. In particular, they lack the ability to describe behaviour and semantics descriptions required to ensure interoperability.

An information/data model is built according to specific necessities and it is made up by the network operator following a known technology orientation, where the interoperability is not considered as a critical requirement most of the time. SOA aim to cope with part of those necessities focusing mainly in end-user requirement and developing the software algorithms as mechanism to achieve them [Sedaghat11]. However, even SOA lack in power to bridge the gap between technology and service requirements.

Cloud computing emerges as a requirement to satisfy the lackness of interconnectivity between physical infrastructure and software services demands. Cloud computing refers to the on-demand provisioning of shared and distributed computing services or infrastructure over the Internet. The cloud computing paradigm

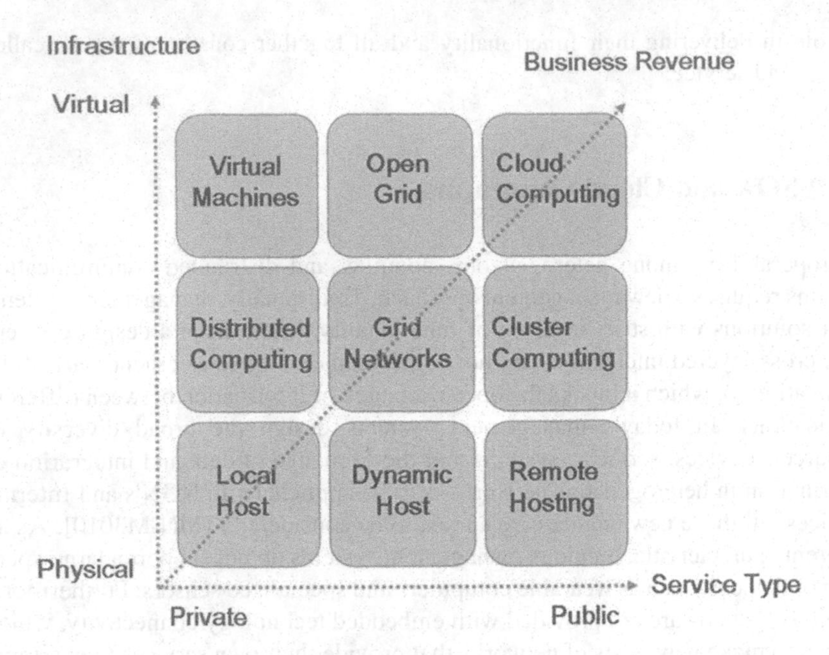

Fig. 1.2 The three dimension in cloud computing

offers one of the largest and most powerful concepts about service provisioning in the ICT domain by using a shared infrastructure [MicrosoftPress11]. Shared by means of multiple service providers interacting to support a common service goal and distributed by means of multiple computers on which to run computer applications by providing a service, mostly Web services. Allocating services in the cloud facilitates and simplifies computing tasks and reduces price for operations, promoting the pay-as-you-go usage of computing services and infrastructure [Bearden96].

In particular, the cloud computing paradigm relies on the business objectives of secure and reliable outsourcing of operations and services, the local or remote infrastructure and the infrastructure type to define its perfect model of best revenue with less technological investment. The three dimension in cloud computing is shown in Fig. 1.2. Unlike conventional proprietary server solutions, cloud computing facilitates, in terms of time and resources, flexible configuration and elastic on-demand expansion or contraction of resources [Domingues03].

So enterprises interested on utilizing services in the cloud have come to realize that cloud computing features can help them to expand their services efficiently and also to improve the overall performance of their current systems by building up an overlay support system to increase service availability, task prioritization and service load distribution, and all of them based on users' particular interests and priorities.

1.5 Conclusions

This chapter discusses alternatives to facilitate the interoperability of the information in management systems by semantically enriching the information models to contain additional references in the form of semantic relationships to necessary network or devices concepts, defined in one or more linked data descriptions. Then, systems using information contained in the model can access and do operations and functions for which they were designed.

This chapter discuss about augmenting management information described in information and data models with ontological data to provide an extensible, reusable common manageability layer that provides new tools to better manage resources, devices, networks, systems and services. Using a single information model prevents different data models from defining the same concept in conflicting ways.

This chapter introduced cloud computing, presenting a brief discussion about the concepts, services, limitations, management aspects and some of the misconceptions associated with the cloud computing paradigm. This section addressed advances of state of the art in the cloud computing area.

Chapter 2
The Role of Semantics in ICT at the Cloud Era

2.1 Introduction

This chapter introduces overall foundations to identify the role of knowledge engineering area and particularly semantic aggregation activity in the ICT and communication areas and their influence for a possible convergence to support communication services. The described foundations guide you for constructing a conceptual framework around a particular problem in managing communication services which is endowed to the information used for management purposes. The semantic expressiveness is necessary for supporting Internet services, allows the customization, definition, deployment, execution and maintenance of services, and then improves the management functions and operations in multiple networks and service applications. The role of semantic applications which support these services can be used in ICT and telecommunications services where context-aware and autonomic features are also discussed.

Particularly, this chapter defines the basic concepts and requirements that pervasive service applications impose as requirements on network and services infrastructures from a management point of view. This chapter makes references to a wide range of context integration research, followed by an analysis of policy-based management (PBM) models [IST-CONTEXT] and cloud services and infrastructures.

The organization of this chapter is as follows. Section 2.1 presents the importance of using ontologies in pervasive services and management operations, in a way of initial description for understanding the importance of semantics in the cloud era.

Section 2.2 introduces fundamental concepts and terminology used. The definitions of terms related to diverse domains, such as context-awareness and context modelling, service management and operations, ontology engineering and autonomic communications, virtual infrastructures and cloud computing, with the objective of building the conceptual platform are used in this book.

Section 2.3 addresses the need to understand and recognize ontology, the importance of providing semantic-oriented and pervasive solutions, and highlights why

J.M. Serrano Orozco, *Applied Ontology Engineering in Cloud Services, Networks and Management Systems*, DOI 10.1007/978-1-4614-2236-5_2,

the integration of context information, with formal semantic mechanisms as ontologies, in service operations to support network and service management, is relevant for cloud services management. At the same time and briefly this section is intended to present trends in pervasive management to understand how semantics are being used in autonomic communications and cloud computing.

Section 2.4 discusses the semantic aspects of knowledge engineering and context information in the form of ontology operations and communications services supporting cloud computing. Challenges regarding context information, information models, service management and ontologies are discussed in terms of what, why, where and most importantly, how those concepts are part of the cloud computing challenges.

This section acts as the state of the art for the requirements in current IT infrastructures and cloud service management. This section introduces the major areas related to information requirements, end-user requirements and technological requirements, and discusses cloud computing challenges and trends. Commonalities are highlighted to produce a compilation or research challenges list and technological requirements.

Section 2.5 introduces the chapter conclusions in form of a discussion about concepts and ideas from the diverse challenges and trends in cloud computing introduced in this chapter.

2.1.1 Importance of Using Ontologies in Pervasive Services and Management Operations

In communication systems, pervasive computing (frequently named ubiquitous computing indistinctly), plays an important role when information processing and management is required. First because it has an inherent feature and/or ability to take advantage of collected pieces of information and related them with service applications and second because in pervasive computing the context-awareness is considered as the main aspect at the beginning of the system design. The advantages that context-awareness brings are principally in the automation of operations (e.g. in the support of activities related to self-management operations in communication networks), when new context information can trigger the deployment of new services improving their performance or developing new services or applications for end users.

It is clear that the complexity for designing and deploying support systems (i.e. customization, management and billing) for context-aware services is also high and requires appropriate tools and adequate infrastructures. One of the most difficult aspects in managing context-aware services is gathering the context information. There are many different proposed standard formats of context information, including people profile, location, network properties or status, applications and other application-specific information as example. This poses the challenge of tracking different mechanisms to gather the information and process it in real time, and then react according to those changes. Another and more important challenge is the modelling and structure of the context information. Without a model, applications will not be able to use such information. Nevertheless, this model must be rich and flexible enough to accommodate not only the current but also future aspects of context information. Another important aspect to consider is that the model should be based on standards as much as possible and it should scale well with the network or the application.

The clear challenge is how modelling and structuring of the context information to gather, distribute and store the data. Context information in the framework of this book refers to operations in management functions. Within the vision of this book, with information models to represent and for integration of management operations and pervasive applications, ontologies play the role of those formal mechanisms that are able to model and use such information in a way that management operations come up as a result of more flexible and efficient creation, adaptation and deployment of communications services. Ontologies are essentially the tool to enhance, semantically, pieces of information and make them accessible to multiple and diverse applications.

The ontology-based model representing context information model must be self-describing, flexible and formal enough for representing not only the current status of the managed object, but also the future aspects that could be defined later as context information. The model should scale well with the network or the application domain. Most current applications use data models (and sometimes, though rarely, information models) for defining context information. An information model is usually not used, since these models are adapted for specific applications, and do not provide enough descriptive and semantic enrichment to model the interaction between applications at different abstraction levels to support cross-layered, interoperable operation.

2.2 Important Concepts and Terminology

The set of definitions, as they are being used in this book, are presented in this section.

2.2.1 Context Information

As an introductory premise, it is necessary to know that context information has been under research for long time [Dey97]. One of the most popular definitions of context introduces the concept entity and the interactions between them [Dey00b]. Due to its dynamicity and huge variety, context information cannot be defined as static information, so then it is not difficult to imagine the multiplicity of definitions aiming to capture part of the context features. Normally context definitions are associated with the application where the information is used and processed, rather than with the own context concept [Bauer03], [Brown97]. In the framework of this book, in order to have a common understanding about context and its application in communications, the concept is introduced considering the existing cited definitions and the context information can be related to entities and the entities can be handling as objects with its respective properties. As a simple concept, the most useful definition is aligned with the most recent context definition, "*the context of an entity is a collection of measured and inferred knowledge that describe the state and environment in which an entity exists or have existed*" [Strassner08].

2.2.2 Context-Awareness/Ubiquitous Computing

Applications taking advantage of environmental information, principally outdoor locations, can be considered as the precursors of the concept of context-awareness [Abowd97]. In fact, the conceptual definition has not evolved too much. In the framework of this book: "*context-awareness is the capability that helps certain applications to exhibit a more personal degree of interaction with the user*" [Abowd97], [Dey97]. More specifically, the certain applications are intending for helping to solve different technological service problems and particularly management problems.

2.2.3 Policy-Based Management

In the field of network management, a policy has been defined as a rule directive that manages and contains the guidelines for how different network and resource elements should behave when some conditions are met [IETF-RFC3198]. In other words, a policy is a directive that is specified to manage certain aspects of desirable or needed behaviour resulting from the interactions of users, applications and existing resources or services [Verma00].

In the framework of this book, an initial definition to consider is "*Policy is a set of rules that are used to manage and control the changing and/or maintaining of the state of one or more managed objects*" [Strassner04], since this definition is more applicable to managing pervasive services and applications. The inclusion of state is important for pervasive systems, as state is the means by which the management systems knows if its goals have been achieved and if the changes that are being made are helping or not.

2.2.4 Pervasive Services

A pervasive service has been defined as a service that takes into account part of the information related to context in order to be offered [Dey00a]. However, as a result of the advent of new devices that are faster, more efficient, and possess greater processing capabilities, pervasive services have been conceptualized as more device-oriented, and the applications designed for them consider not only single user benefits, but more importantly, the interaction between users and systems. This increases the effect that mobility has on a pervasive service. In the framework of this book, a *pervasive service is one that makes use of advanced ITC mechanisms to facilitate service management operations and manifest itself as always available.*

2.2.5 *Ontology Engineering*

In the framework of this book, ontology is a specification of a set of concepts using a formal language that can define and describe properties, features or relationships between the concepts in a particular domain or even between concepts of different domains (e.g. management networks and services).

The term ontology is often used in a more generic way, as a formal and explicit specification of shared conceptualizations [Gruber93b], [Gruber95]. In other words, ontology is a formal description of the concepts and relationships that can exist for an agent or a community of agents [Guarino95].

This definition is consistent with the usage of ontology in communications, thus an ontology is a standard definition or array of concepts. In terms of a standard/formal language, ontology is a conceptualization representing abstract model(s) that contain concepts that are relevant to a particular domain, management operations of pervasive and autonomic communication domains, virtual infrastructures and/or cloud computing, for example.

2.2.6 *Autonomic Communications*

This section discusses the use of autonomic communications, highlighting the management of information and resources, service re-configurability and deployment, and the self-management requirements inherent in autonomic systems. The purpose of autonomic systems is to solve problems of managing complex service and communications systems [Strassner06c], [IBM05].

Autonomic systems are the result of information technologies interacting and cooperating between them for supporting service management operations (e.g. creation, authoring, customization, deployment and execution).

The interaction is supported by the extra semantics available from context information using ontologies and other information technologies, such as the policy-based paradigm for managing services and networks [Serrano06b]. The interactions are shown in Fig. 2.1.

Autonomic computing is the next step towards increasing self-management in systems and coping with the complexity, heterogeneity, dynamicity and adaptability

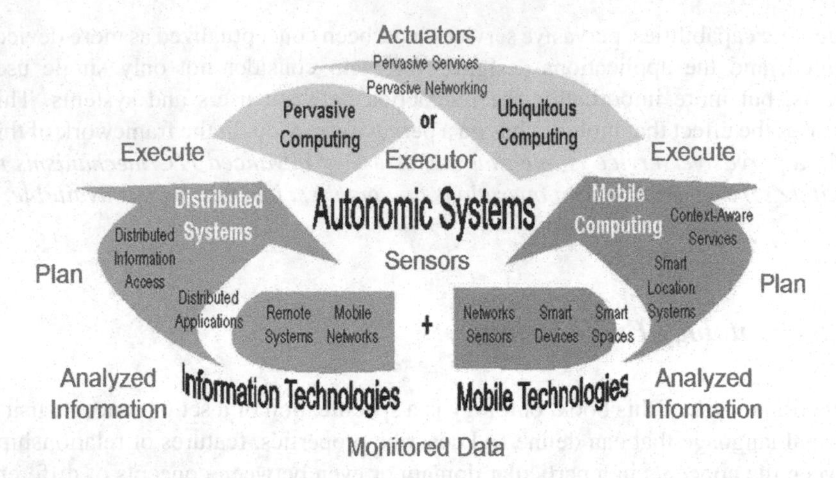

Fig. 2.1 Combining autonomic computing with communication systems and technologies related

required in the modern communication services systems. This trend of increasing complexity is evidenced by the convergence of advanced pervasive systems with engineering technologies (networking, service and knowledge and information technologies) and the Internet capabilities.

Autonomic communications support the idea of autonomic elements, which process the management information to provide the autonomy necessary to support next-generation networks and services.

This section argues that context, which provides increased semantics (and hence, increases the ability to reason about management data), should be used to augment existing management to fulfil the goal of providing self-managing functionality, this is a premise used to understand the autonomic concept in this book. Autonomic elements are to be built to satisfy information models; hence, context should be added as an element of the information model (or at least made available to it). This will not only provide greater functionality, but also enhance the interoperability of the system.

Autonomic services, as much as systems and networks become more self-managing, the nature of the services should evolve accordingly. Thus, the services should be developed based on autonomic computing principles, so that the services also become self-managing.

Autonomic elements, services must be more flexible in order to respond to highly dynamic computing environments and be more autonomous to satisfy the growing and changing requirements from users. This means that services and components (software pieces or devices) need to be more context-aware. The addition of context to information models, augmented by ontological information, is the key to creating services that are self-managing.

Autonomic management, to cope with increasing management complexity, it is necessary that systems support the eight features that characterize an autonomic system, as defined by IBM [IBM01b]. These eight features have guided the development of autonomic systems. The following subsections review these critical autonomic communications features [IBM01b], [Strassner06c].

2.2.6.1 Self-Awareness

As per definition, self-awareness is the capability of a system to collect pieces of information and process them locally, this is a feature to be aware of its state and its behaviour, with the objective of governing operations itself, and for sharing and collaborating its resources with other systems in a more efficient manner.

2.2.6.2 Self-Configuring

This is the capability of a system for adapting dynamically and automatically to changes in its environmental conditions. Self-configuring can refer firstly to configuring conditions (setup), which must be done largely automatically in order to handle changes in the environment, and secondly to the adaptability of the architecture to re-configure the system for achieving service quality and experience factors.

2.2.6.3 Self-Optimization

This is the capability of a system to improve the resource utilization and workload of the system following the requirements from different services and users. This resource utilization and workload is dependent on the time and service lifecycles for each service and user. Performance monitoring and resources control and optimization are management operations inherent in this process.

2.2.6.4 Self-Healing

This is the capability of a system to detect and prevent problems or potential problems, and then, as a result of this action, to find alternate ways of using resources or reconfiguring the system to avoid system or service interruptions. To perform this capability, local data processing is needed.

2.2.6.5 Self-Protection

This is the capability of a system that defines its ability to anticipate, detect, and protect intrusion or attacks from anywhere. This depends on its ability to identify failures in the system, and enables the system to consistently enforce privacy rules and security policies.

2.2.6.6 Context-Awareness

This is the capability of a system to process pieces of information in their environment, including their surrounding, and activity, with the objective to react to that information changes in an autonomous manner as a result of the utilization of elements and process in its environment.

2.2.6.7 Open

This is the characteristic of a system to operate in heterogeneous environments and across multiple platforms, always using and implementing open standards. Autonomic systems must operate in a heterogeneous world, and hence cannot be implemented using proprietary solutions.

2.2.6.8 Anticipatory

This is the characteristic of a system to be prepared for providing the optimized resources needed in order to seamless provide better functionality for its users, all the while keeping the complexity of the system hidden.

It is important to highlight, in this book, that these eight main features for autonomic systems have been related to the most important generic service operation requirements to determine the impact that these features has on service requirements, and to create a vision in how these service operation characteristics can meet the requirement of an emerging areas such as cloud computing.

Figure 2.2 shows that context-awareness is one of the most important features to be considered when supporting services are related to all of the other service operation requirements [Serrano06b]. Thus, its crucial handling and dissemination of the context information support the pervasive services and its feature of context-awareness enables autonomic systems control. Self-configuring is also a requirement that is related to all of the other service operation requirements; however, it is related more to the implementation perspective.

2.2.7 Virtual Infrastructure and Cloud Computing

It is anticipated that cloud computing should reduce cost and time of computing and operations processing [IBM08]. However, while cost benefit is reflected to end user only, from a cloud service provider perspective, cloud computing is more than a simple arrangement of mostly virtual servers, offering the potential of tailored service and theoretically infinite expansion [Head10]. Such a potentially large number of tailored resources which are interacting to facilitate the deployment, adaptation and support of services, this situation represents significant management challenges.

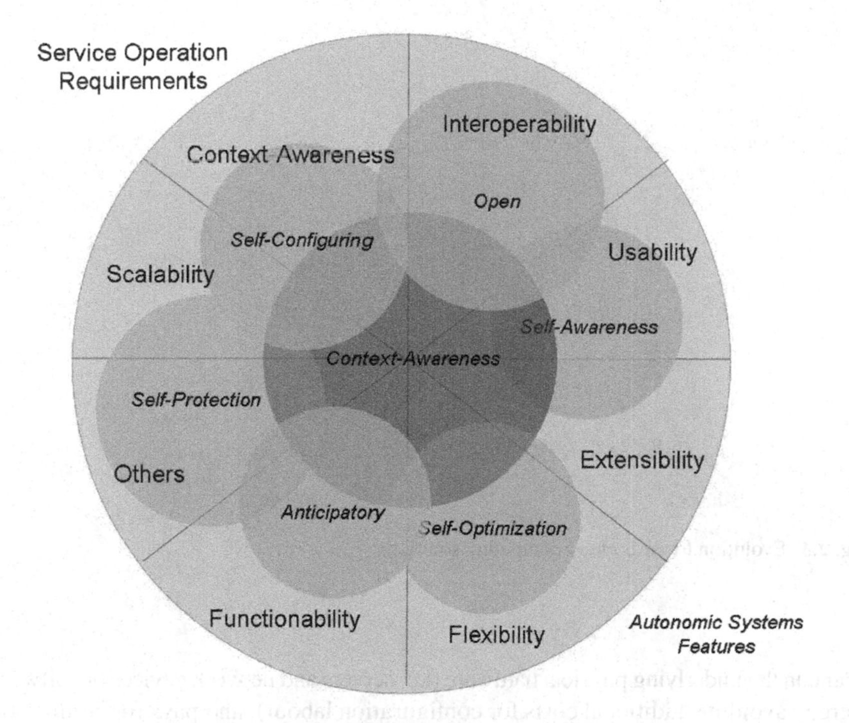

Fig. 2.2 Relationships between autonomic systems features and service operation requirements

In management terms, there is a potential trend to adopt, refine and test traditional management methods to exploit, optimize and automate the management operations of cloud computing infrastructures [Waller11]; however, this is difficult to implement, so designs for management by using new methodologies, techniques and paradigms mainly those related with security are to be investigated.

The evolution of cloud computing has been benchmarked by a well-known evolution in distributed computing systems [IFIP-MNDSWG]. Figure 2.3 depicts this evolution and shows the cloud computing trends passing from a physical to virtual infrastructure usage and from a local to remote computing operations. This evolution towards cloud computing services era is briefly explained in the following sections.

2.2.7.1 Virtualization

Cloud computing is leading the proliferation of new services in the ICT market, where its major success has been to facilitate on-demand service provisioning and enabling the so-called pay-as-you-go provision and usage of resources and server time over the Internet. The user of the cloud services does not own and does not

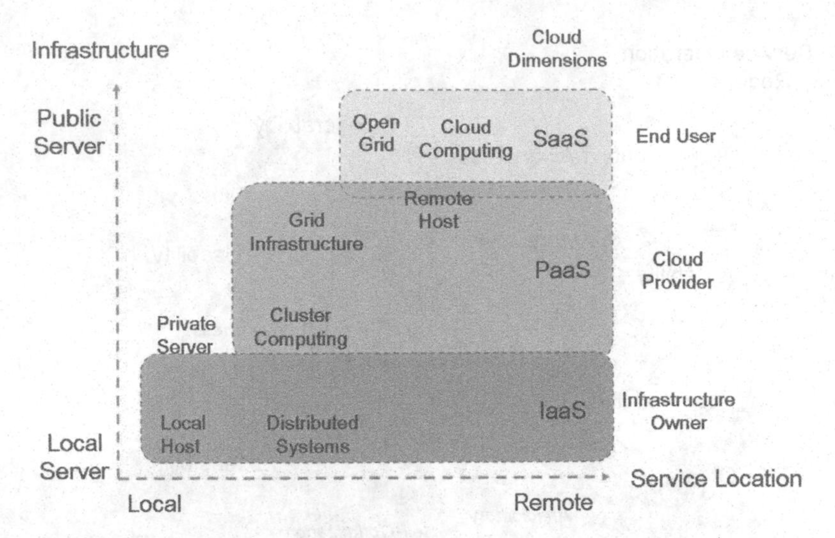

Fig. 2.3 Evolution towards cloud computing solutions

maintain the underlying physical hardware (i.e. servers and network devices or software thereby avoiding additional costs for configuration labour), and pays for permission to use "a virtual slice" of those shared resources.

As today's cloud computing is known, it has origins in distributed computing systems [Andrews00], [Elmasri00], [Lynch96]. An important feature in distributed systems is the role management systems have in order to control, processes remotely and in a coordinated manner. After a decade of management development, and evolution in computing systems emerges grid computing, a combination of remote computer resources to execute common goals in remote located physical infrastructures [Foster99]. In grid computing no matter where the resources are allocated, tasks are executed in distributed places by using grid managers to pursue multiple tasks even through different administrative domains [Catlett92], [Maozhen05], [Plazczak06], [Buyya09].

In generic terms, grid computing can be seen as a distributed system where large number of non-interactive files is involved. It is important to mention what make grid computing different from cluster computing is that grids seem more seamless coupled, it is definitively heterogeneous, and geographically their allocation is spread out. Grid computing is traditionally dedicated to a specialized application and it is a more common cluster computing which will be used for a variety of different purposes. Likewise grids are more often build with the aim to be used in a more long-term application and involve more use of physical infrastructure resources with specialized or particular ad hoc developed software libraries known as middleware. Examples of middleware are GridWay [GRIDWAY], TeraGrid [TERAGRID], gLite [GLITE], UNICORE [UNICORE] and Globus Toolkit [GLOBUS].

2.2.7.2 Cloud Service

As main feature in cloud computing systems, while labour costs are still present, users pay reduced prices as the infrastructure is offered to and shared by multiple users [Head10], [Urgaonkar10]. According to this model, processing time cost for each user is reduced since it is covered by multiple users, as seen in the traditional pay-for-server-time model [Greenberg09].

It is well accepted by ITC professional that cloud computing is a revolution in the service provisioning and marketing giving an opportunity to bring to bear technological experience and revenue in a new area by exploiting the Internet infrastructure. However, the concept behind this trend is the full exploitation of multitenancy of services, where multiple users can make use of the same infrastructure by using intermediate middleware's known as virtual infrastructure to use the same information service [Greenberg09].

2.3 Ontology Engineering, Autonomic Computing and Cloud Computing Basis

In telecommunications, pervasive computing applications have always attracted the attention of research and industry communities, mainly because pervasive computing conceptually offers to all of its users (at any level of abstraction, from end user to network operators) broad possibilities for creating/supporting useful and more efficient services.

Pervasive computing requires a mixture of many technologies and software tool/applications to materialize these benefits [Serrano06d], as shown in Fig. 2.4; however, it is context-awareness, the feature that makes the computing applications in pervasive

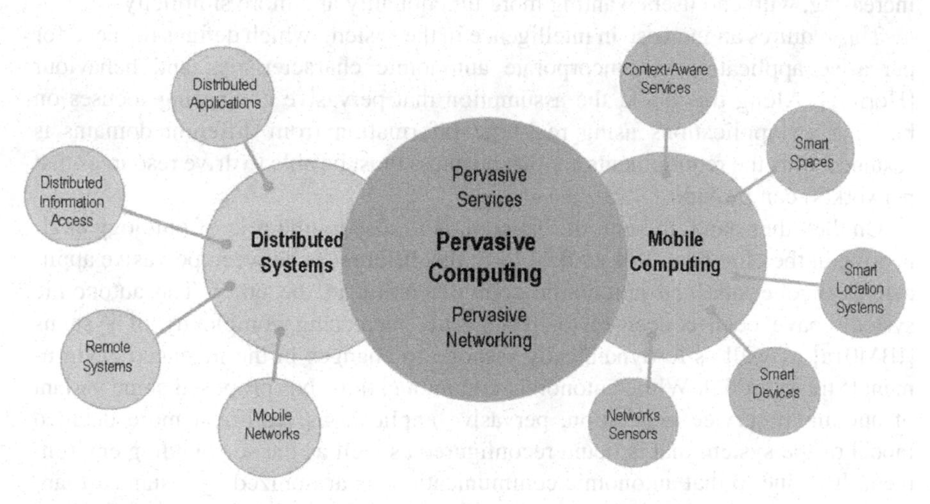

Fig. 2.4 Pervasive computing composition

services more extensible, useful and many times simple. To make this more attractive, this trend will continue, as new mobile and wireless technologies are being integrated.

It is not difficult to imagine that this mixture of technologies increases the complexity of the systems and solutions by many systems and devices that use different mechanisms for generating, sharing and transferring information to each other. Further more existing data models are application-specific and designed independently to each other. In this respect, a method or set of them are needed to enable the efficient and clear exchange and reuse of information between the systems.

This method must be inherently extensible; as systems and networks become more pervasive, the nature of the services provided should be easy to change according to changing context. Services must also become more flexible in order to respond to highly dynamic computing environments and become more autonomous to satisfy the growing and changing requirements from users. In other words, the services must become more adaptive and context-aware.

Simple advances in resources and services have been feasible as a result of the ever-increasing power and growth of associated technologies. However, this drive for more functionality has dramatically increased the complexity of systems—so much that it is now impossible for a human to visualize, much less manage, all of the different operational scenarios that are possible in today's complex systems.

The stovepipe systems that are currently common in OSS (operations support system) and BSS (business support system) designs exemplify this—their desire to incorporate best of breed functionality prohibits the sharing and reuse of common data, and point out the inability of current management systems to address the increase in operational, system, and business complexity [Strassner06b].

Operational and system complexity are induced by the exploitation and introduction of technology to build functionality. The price that has been paid is the increased complexity of system installation, maintenance, (re)configuration and tuning, complicating the administration and usage of the system. Business complexity is also increasing, with end users wanting more functionality and more simplicity.

This requires an increase in intelligence in the system, which defines the need for pervasive applications to incorporate autonomic characteristics and behaviour [Horn01]. Along this book, the assumption that pervasive computing focuses on building up applications using real-time information from different domains is assumed, thus the requirements are that business must be able to drive resources that network(s) can provide.

On the other hand, the aim of this book is to discuss the role of ontology engineering in the cloud era. It is evident from the differences between pervasive applications in one side and autonomic communication in the other. The autonomic systems have been conceived to manage the increasing complexity of systems [IBM01a] as well as to dynamically respond to changes in the managed environment [Strassner06a]. While autonomic communications has proposed some variant of automatic service generation, pervasive applications require a more detailed model of the system that is being reconfigured as well as the surrounding environment. It is due to that autonomic communications is armonized by a standard language where the sharing of information is easier than when formal but different languages are being used as it occurs in pervasive applications.

More importantly, a system for supporting pervasive applications cannot be reconfigured if the system does not "understand" the functionality of the components and any restrictions imposed on that functionality by its users or the environment. This means that the system requires self-knowledge of its components, and knowledge of its users and requirements. In pervasive systems one of the most important forms of knowledge and it is most basic stage is context, as it can be used to simply capture knowledge.

An example aiming to clarify the relationship between pervasive applications and autonomic communications is given as follows: let us assume that a pervasive application is able to generate a code to modify performance itself and this is based on variations of contextual information. The model of an autonomic system using context information and its relations is not just the set of self-configuration and other self-* operations; self-knowledge and self-awareness are required in order to provide autonomic behaviour, with the resulting set of decisions based on business rules as well as context-specific information. Since functionality can change, the autonomic solution must incorporate extensible information and data models to accommodate future decision and operations.

2.3.1 Ontologies to Define, Represent and Integrate Context Information

The nature of the information requires a format to represent and express the concepts related to the information. Ontology is an explicit and formal way to capture and integrate information, without ambiguity, so that the information can be reused and shared to achieve interoperability. Explicit means that the types of concepts used, and the constraints on their use, are unambiguously defined. Formal indicates that the specification should be machine readable and computable.

A specific definition of "Ontology," with respect to system and network management, is a database describing the concepts in a domain, their properties and how the concepts relate to each other. This is slightly different from its definition in philosophy, in which ontology is a systematic explanation of the existence of a concept. System and network management is less concerned with proving the existence of something than in understanding what that entity is, and how it interacts with other entities in the domain [Guarino95].

2.3.2 Using Ontologies for Managing Operations in Pervasive Services

Pervasive computing or ubiquitous computing has evolved and acquired a grade of pervasiveness such that pervasive computing promises to be immersed in everyday activities and environments. Pervasive computing goes beyond the idea that almost any device, from personal accessories to everyday things, such as clothing, can have embedded computers that can create connections with other devices and networks. The goal of pervasive computing, which combines current advanced electronics with network technologies, wireless computing, voice recognition, Internet capabilities and artificial intelligence, is to create an environment where the connectivity of devices and the information that they provide is always available.

However, in this complex environment where systems are exchanging information transparently using diverse technologies and mechanisms, management becomes increasingly difficult. The increasing multiplicity of computer systems, with the inclusion of mobile computing devices, and with the combination of different networking technologies like WLAN, cellular phone networks, and mobile ad hoc networks, makes even the typical management activity difficult and almost impossible to be done by human beings. Thus, new management techniques and mechanisms must be applied to manage pervasive services.

One of the most important characteristics of knowledge is the ability to share and reuse it. In this context, ontology is used for making ontological *commitments*. An ontological commitment is an agreement to use a vocabulary (i.e. ask queries and make assertions) in a way that is consistent. In other words, it represents the best mapping between the terms in ontology and their meanings. Hence, ontologies can be combined and/or related to each other by defining a set of mappings that define precisely and unambiguously how concepts in one ontology are related to concepts in another ontology. Thus, ontologies are a powerful means to provide the semantic structures necessary to define and represent context information.

Ontologies were created to share and reuse knowledge in an interoperable manner [Guarino95] and to avoid the handicaps founded when different systems try to exchange application-specific heterogeneous representations of knowledge. Such cases are complicated by the difficulties between languages and patois; missing of communication conventions and mismatch of information models.

As a data or information model represents the structure and organization of the data elements, the management activity can obtain benefits from using the data from such elements in its operations. In principle, a data or information model is specific to the application(s) for which it has been used or created. Therefore, the conceptualization and the vocabulary of a data model are not intended a priori to be shared by other applications [DeBruijn03]. Data models, such as databases or XML schemas, typically specify the structure and the integrity of data sets. The semantics of data models often constitute an informal agreement between the developers and the users of such data, and that finds its way into applications that use the data model. By contrast, in the area of knowledge engineering, the semantics of data needs to be standardized in a formal way in order to exchange the data in an interoperable manner [Genesereth91].

Ontologies not only provide enrichment to the information model, but also semantic expressiveness, allowing information exchange between management applications and different management levels. It is this characteristic by which ontologies are emerging into engineering areas, providing advantages to better specify the behaviour of services and management operations. One potential drawback of

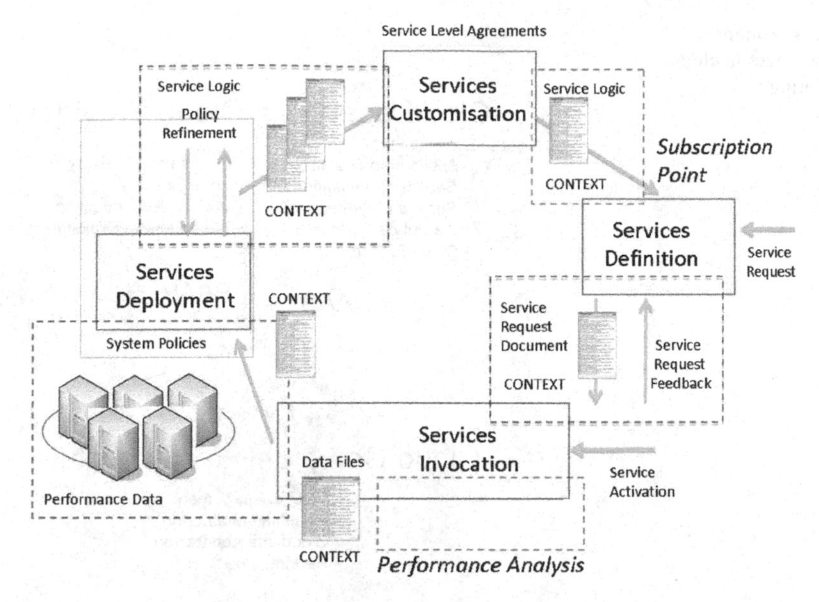

Fig. 2.5 Context information role in pervasive computing environments

using ontologies for management purposes is that they require significant computational resources. However, the disadvantages are outweighed by the associated benefits of using ontologies. Times for agreements are reduced when a system using ontologies is being used, for instance or even more when the information into the systems need to be shared to other systems, the times for seeking of the information and mapping are reduced.

2.3.3 *Autonomic Communications and Ontologies*

Autonomic systems emerge as one way to solve management complexity. Complex interactions need to be supported by the increase in semantics embedded in the context information that is represented by ontologies and controlled using policy languages. Autonomic communications can be seen as an approach consisting of the integration of context information in management operations for pervasive services that can be extended into changes in the communication networks by autonomic element executed in the network infrastructures. Specifically, autonomic communications focus on the integration of context information to control management operations in the service lifecycle; to achieve this goal, use the ontology descriptions for management operations and formal modelling techniques.

Figure 2.5 depicts an scenario in which the level of complexity is easily visualized, the context information contained in the networks and devices can be used for multiple operations, even in different domains, and give support for management

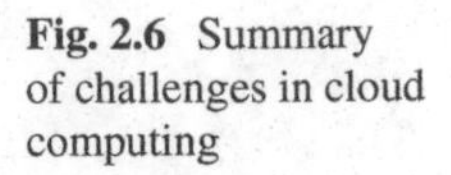

Fig. 2.6 Summary of challenges in cloud computing

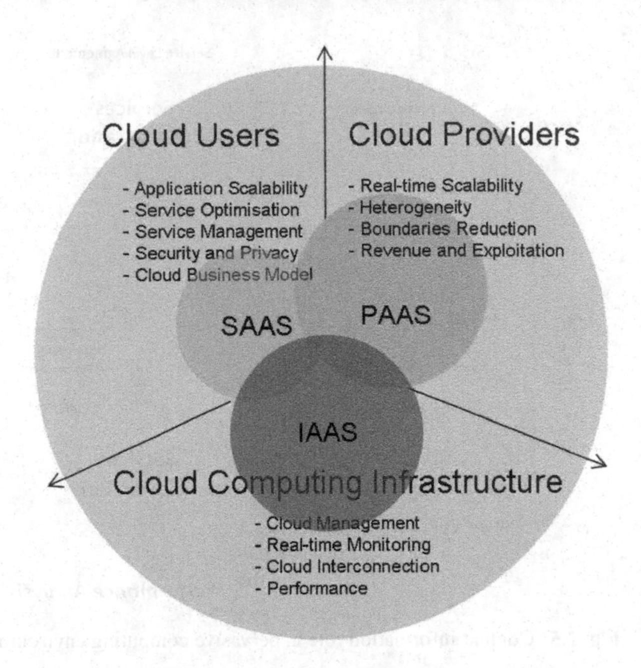

operations such as customization, definition and deployment and, even more importantly, the service maintenance. In the depicted autonomic environment, the possibility of upload or transfer of context information from its various sources to the management system increases the pervasive level of the applications and the services using such information (shown on the left side of Fig. 2.4).

This level is depicted as a bar in order to indicate that context information must be *translated* between each level. The translation is simplified and automated when formal languages are used. One of the objectives of this book is to explain the need for making context network information available to different service abstraction layers, such as those shown in Fig. 2.4 for triggering appropriate management operations in autonomic systems and also to provide guidance to find out the best possible approaches to achieve this goal.

2.3.4 Cloud Service Challenges and Ontology Engineering

In this section, a number of research challenges for cloud computing is described. It is not intended to be an exhaustive list, but rather to introduce important issues with respect to improving cloud management, and which are yet open issues and under investigation in the emerging cloud computing area. Figure 2.6 summarizes these trends based on well-identified cloud demands [Greenberg09], [DEVCENTRAL].

Figure 2.6 depicts trends and challenges in cloud computing, particularly from a user perspective, where cloud computing is being focused according to IDC

Enterprise Panel, "clients themselves are concerned with Security, Availability and Integration, today, and over the next 3 years, customers will be moving to Cloud Services" [DEVCENTRAL].

2.3.4.1 Real-Time Scalability of the Cloud Service

Cloud services must reflect the current total service load to address time-varying demand. Thus, the application must actively participate in the process to produce changes in the underlying virtual or physical infrastructure. In other words, individual services need not to be adaptive with system load; however, they generate information that can be used as inputs for applications used to modify infrastructure performance. Therefore, the application is not static anymore; in the cloud, it can evolve over time to migrate or expand using more or less computing resources. So cloud services are agnostic of the underlying infrastructure, where possible, and it is the job of the physical infrastructure to host the services in a self-adaptive manner.

2.3.4.2 Scalability of the Application

In cloud computing, adding more resources does not necessarily mean that the performance of the application providing the service will increase accordingly. In some situations, the performance may decrease because of the managing/signalling overhead or bottlenecks. A challenge is to design proper scalable application architectures as well as making performance predictions about data and processing load balancing to satisfy service load demands and prevent future demands.

2.3.4.3 Optimization for Complex Infrastructure and Complex Pricing Model

If well-known pay-as-you-go model makes cloud very attractive, there are many possible other models for application architectures where revenue is the main benefit. A challenge is to design an optimal architecture to fit a cloud environment and at the same time to optimally use all resources with respect to pricing and billing. Note that the pricing model may conflict with the optimally scalable architecture, thus requiring a carefully designed trade-off of application and infrastructure design priorities is the challenge.

2.3.4.4 Cloud Scalability Limitations

Cloud services are commonly referred to as "infinitely" scalable. However, far away from the technology is near to reach this concept. A concept of this dimensions is not generally possible to solve physically by just expanding the server farm, many systems working cooperating and co-ordinately are needed. It is of interest to enable federation of resources from several cloud providers. This has recently been addressed and emphasized by Google through a statement made by Vint Cerf at the Churchill Club on FORA.tv on the Inercloud and the Future of Computing [FORATV].

2.3.4.5 Maintaining Service Levels

As in networking services, with increasing amount of resources consumed, reduced service levels and failures are commonplace, in other terms the quality of the service is specifically in the case of IaaS, when applications must handle them without any interruption. Networking resources are seen as a particularly critical resource and a common point of failure; hence, QoS is closely related.

2.3.4.6 Security

With the participation of multiple users in the cloud environment, the privacy and security of the information being stored, transmitted or operated is crucial. In public clouds, this issue is still driving research activity in terms of efficiency, data protection algorithms and protocols [Mace11]. This is motivated mainly by legal issues about security requirements for keeping data and applications operating in a particular legal domain (e.g. Ireland, EU, etc.) and not due to technological constraints.

2.3.4.7 Heterogeneity of Cloud Platforms

There are multiple vendor-specific platforms—the possibility to build applications or services that are provider-neutral to prevent lock-in would allow redundant deployments, reduce staff training costs (and OPEX in general), and would also enable best of breed (or cost optimal) service provider selection. Close relation exists between the development of adequate and efficient management tools and/or systems managing the multiplicity [Rochwerger11]. Up-to-date cloud service provisioning has been the central activity around this area; however, once the cloud will achieve a certain level of maturity, the development of inter-connected platforms for the cloud will be a necessity.

2.3.4.8 Management of Underlying Resources—Transparency

In cloud computing, where underlying resources are hidden/abstracted, service management remains like an area where research efforts are required to find the best methods and mechanisms to enable services to be managed externally.

Today the cloud and its underlying infrastructure is managed in the form that services capabilities are not exposed [Srikanth09]. The cloud computing paradigm establishes that computing operations and service applications can be performed at unlimited scalable levels and following on-demand service attention, where (at least from a service end-user perspective) the less important is to worry about infrastructure. However, in PaaS and IaaS levels, a key feature for the cloud optimization and at the same time increase the performance of the cloud infrastructure is that the low-level management of underlying resources and services remains hidden from the application, and the end user has to relieve the platforms and management user, when it is needed, to manage them. However, the end user still requires some flexibility in terms of the resources and service they pay for so [Sedaghat11]. A key challenge remains in the mapping of high-level user requirements to low-level configuration actions. It is also vitally important that these management actions can be performed in a scalable manner since the cloud service will not be able to individually configure a large number of shared resources and services for individual users.

2.3.4.9 Management Scalability Towards Federation

In cloud computing, if the number of partitioned virtualized resources (or instances) grows the challenge of how to control resources individually become more apparent. This can be addressed by allowing users to manage their own delegated resources—non-expert resource management. This raises a number of challenges including: support non-expert monitoring and management, managed delegation of management resources and authorities, and auditing [Goiri10]. This would mean that users could monitor their own resources so they can decide when to request more resources and ensure they are getting a near optimal value for the resources they pay for (depending on the payment model used).

2.4 Identifying Requirements for IT Infrastructures and Cloud Service Management

The identification of service requirements, which is between others one of the main research tasks of this book, is presented in this section. This section is divided into four parts: (1) information requirements, (2) end-user requirements, (3) technology requirements and (4) rather to present an exhaustive list of cloud computing challenges and service requirements, it concentrates in the most important up-to-date challenges and where this emerging area has advanced developing tool and technologies to achieve cloud computing level has up to date.

The considerations presented in this section are based on exhaustive study, analysis and discussions about the state of the art in IT and cloud systems and on technical experience from research activities. Basic conceptual frameworks applied developing technologies through participation in international European projects and research network of excellence. Also public documentation that guides the evaluative research tasks to define the state of the art of context-aware services in the pervasive computing knowledge area is available in [IST-CONTEXT], [SFIFAME], [AUTOI], and [IST-EMANICS]. This section acts as a platform research work towards a conceptual framework based on integration to define solid basis and technological principles for supporting pervasive services in cloud environments.

Particularly, pervasive services are often related to mobile devices, because in such cases the location of the device or user, or the type of the user's connectivity, is changing. These factors emphasize one of the most important characteristics of pervasive services: their ability to adapt their service (most likely in as seamless manner as possible) to these changes, and so facilitate truly mobile applications. Pervasive services can also be useful for non-mobile users or users who have a fixed connection or a set of network applications.

For example, the environment of a user may also change and, in such case, the user experience and rapid service re-provisioning can also be enhanced by pervasive applications. Changes in the network conditions can be performed by the pervasive service and the user is insulated from those changes as a result of pervasive service acting.

It is important to emphasize the context-aware capabilities of pervasive services which enable new types of applications in pervasive computing giving the opportunity to built based on heterogeneous technologies, platforms offering inter-connected services. These applications, for example can help users to find their way in unfamiliar areas, receive messages in the most useful and suitable manner, or find compatible people. (A traditional example for pervasive service applications.) All of these and other activities require the system and its devices to be managed in an autonomic manner. This is because the use of context information in pervasive applications reduces the amount of human intervention that an application needs for satisfying user's requests.

The current growing market of middleware applications is also focused on providing applications, from different levels of abstraction (e.g. business, architecture, or even programming points of view), the ability to share information and thereby deliver relevant context information. These middleware applications result in an increased number and development of applications that are able to use context information in order to improve the services that they deliver [Brown97]. In other words, context information is not just user-oriented, as in the beginning when the concept of context-awareness was first introduced, but it is also related to the management of different services and applications. It is only recently that, with the intervention of pervasive computing, more applications using context information have started to be developed as services with context-aware properties. Context-aware applications can enhance the capabilities of devices and applications, and such services are known as pervasive services. For example, in [Abowd97], applications using context information for different tasks in the ubiquitous and services areas are described, while in [Tennenhouse97], there are applications using context for networking purposes.

2.4.1 The Information Requirements

One of the most difficult aspects of context information is its dynamism. Changes in the environment must be detected in real time, and the applications must quickly adapt to such changes [Dey01]. The nature of the information is the most important feature of context-aware applications and systems; if pervasive applications can fully exploit the richness of context information, service management will be dramatically simplified [Brown98]. Other important challenges of pervasive services include: (1) how to represent and standardize the context information, (2) if the information is correctly represented, and (3) if the information can be translated to a standard format, then different applications can all use the information. Finally, some types of context also depend on user interfaces (which can make retrieving context information easier), or the type of technologies used to generate the context information. This sets the stage for discussing the information requirements in this section.

2.4.1.1 Modelling of Information—Context Information Modelling

Modelling context information is one of the major challenges when services deployment design is needed [Krause05]. Without a well defined, clear and at same time flexible information model, applications will not be able to use such information in an efficient way for taking advantage of all of the benefits that the context information can provide for the service as well as for the provisioning of that service. The context information model must be rich and flexible enough to accommodate not only the current facets of context information, but also future requirements and/or changes [Dey01]. It has to be based on standards as much as possible and moreover, the model should scale well with respect to the network and the applications. This introduces a great challenge managing this context information in a consistent and coherent manner. Storage and retrieval of this information is also important. A most well-established approach is to model context information model representation and the framework proposed to represent the context information supported by an object-oriented, entity-centred model.

Figure 2.7 shows an entity-centred model representation. The model is based on simple concepts and its relationships, as syntactical descriptions, between those concepts. An entity is composed of a set of intrinsic characteristics or attributes that define the entity itself, plus a set of relationships with other entities that partially describe how it interacts with those entities.

The entities can represent anything that is relevant to the management domain [Chen76]. Moreover, the relations that can exist between the different model entities can represent many different types of influence, dependence, links and so on, depending mainly on the type of entities that these relationships connect. The model's objective is to describe the entity and its interaction with other entities by describing the data and relationships that are used in as much detail

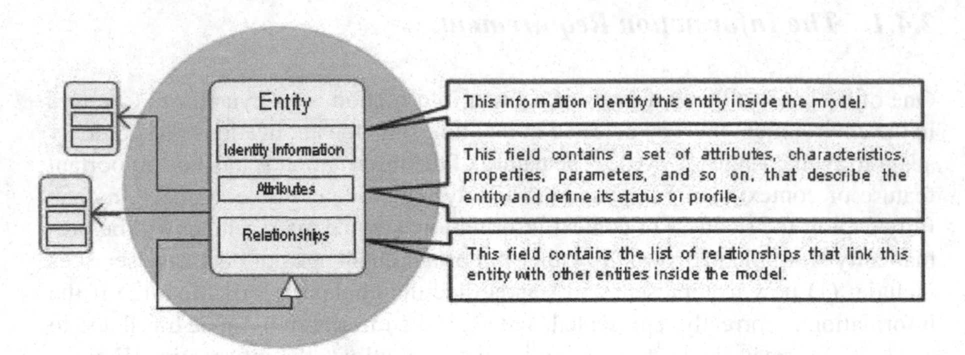

Fig. 2.7 Context model template

as is required. This abstraction enables the model to be made more comprehensible by different applications. Since this format is machine readable, the information can be processed by applications much easier than an equivalent, free-form textual description.

The entity model can be thought as a general purpose way of representing and storing context information throughout the network. Modelling context is a complex task, so using an extensible model provides a "template" that both standardizes the information and enables it to scale its contents for future applications. This entity-centred model can be used to characterize context information.

This model is inherently extensible, and new attributes or relationships can be added to entities without having to modify the structure of the model. The model must be reused, as any new entities that are required can simply be added to the model and linked to the existing entities by defining appropriate (new) relationships, without having to modify any of the existing entities. The entity-centred model is easily scalable. Along this book multiple references which use the entity-relationship paradigm for modelling context information can be found, the reason is simple, by using this model it results in an easy and extensible management mechanism to represent, handle and operate information.

2.4.1.2 Mapping of Information—Context Information Mapping

The entity model provides a powerful abstraction of the information needed by context-aware applications and, in general, the pervasiveness required by such applications. The mapping of context information can be seen as a distributed data base model, where the entities contain only their own information and also the type of relationships with other data entities. This enables an entity to use the attributes of these other entities if needed. This method of representation acts as a suitable scenario description without reference to any specific element or object, and hence is applicable to many different applications. The mapping of the entity model can be thought of as a general purpose way of represent; store and exchange context

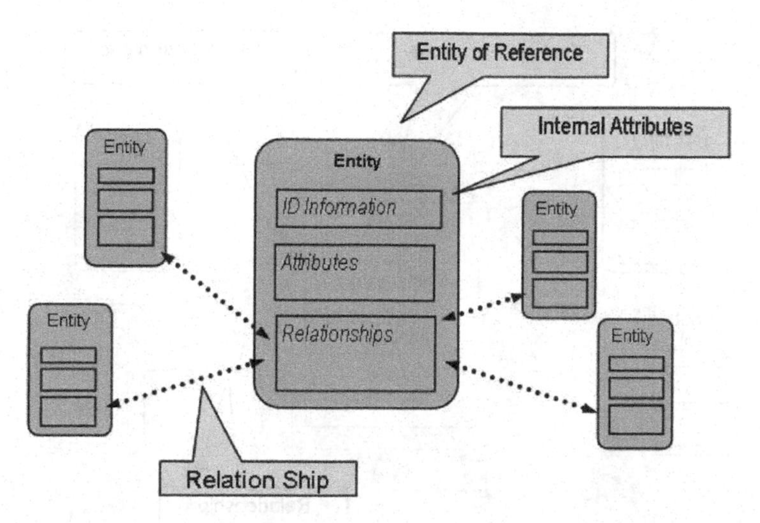

Fig. 2.8 Entity internal architecture

information throughout the network, and for this reason, this concept is used to model context information for mapping purposes.

An abstract model that can be adopted and when necessary add new information to the model is the most appealing solution needed, as a clue for generating this type of models, it is not necessary to modify the existing entities; all that is required is to create a new entity and establish suitable relationships with existing entities. This ensures the scalability of the information model.

Definition of Main Entity Classes

Figure 2.8 shows the architecture of the entity model used and supported in this book. This model defines an object-oriented taxonomy of names required for management, such as Persons, Objects, Places and Tasks. Every entity in the final model will be an instance of one of these generic entity classes. A model for the relationships shall be generated as well, which defines the classes of relationships that must exist (and their attributes), in order to identify different possible relationships between entities and what they mean. The traditional entity definition is described in [Chen76].

Definition of Relationships Classes

The main entities must be related or linked with other entities, of the same or different class, by means of different types of relationships. These relationships also provide useful context information. Since relationships are so powerful, a set of different

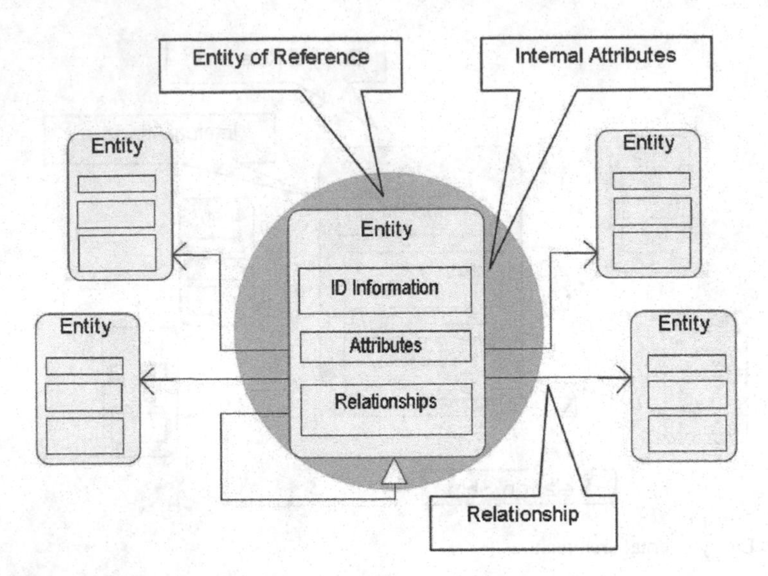

Fig. 2.9 Main relationships between entities

classes of relationships that express the different types of interaction between the different types of entities are defined. Figure 2.9 represents an example set of initial relationships that could be established between the entities of the model. These are high-level type of associations between the four main types of entities. Note that certain types of relationships are only meaningful between certain entities, for example social relations only have meaning between entities of person class and not, for example between places.

2.4.1.3 Taxonomy of Information—Context Information Taxonomy

The classification of context information is not an easy task, due to its extreme heterogeneity. Therefore, this taxonomy could be defined in multiple ways and from multiple perspectives. To identify the information that could be relevant to pervasive applications, the different management operations required by pervasive services were classified based on an extensive literature survey that was conducted to understand the current state of the art in context-aware service provisioning. In this section, a review of different kinds of classifications (most of them orthogonal and compatible) with pervasive services and management operations is presented, a more detailed description can be found in a previous work on [Serrano05]. In a first approximation, context information can be classified by the following characteristics:

By its persistence:

Permanent (no updating needed): Context, which does not evolve in time, that remains constant for the length of its existence (e.g. name, ID card), or Temporary (needs updating): Context information that does not remain constant (e.g. position, health, router interface load).

By its medium:

Physical (measurable): Context information that is tangible, such as geographical position, network resources, and temperature (it is likely that this kind of information will be measured by sensors spread all over the network), or

Immaterial (non-measurable by means of physical magnitudes): Other context information, such as name or hobbies (it is likely that this kind of information will be introduced by the user or customer themselves).

By its relevance to the service:

Necessary: Context information that must be retrieved for a specific service to run properly, or

Optional: Context information which, although it is not necessary, could be useful for better service performance or completeness.

By its temporal characteristics:

Static: Context information that does not change very quickly, such as temperature of a day, or

Dynamic: Context information that changes quickly, such as a person's position who is driving.

By its temporal situation:

Past: Context information that took place in the past, such as an appointment for yesterday, which can be thought of as a context history, or

Present: Context information that describes where an entity is at this particular moment, or

Future: Context information that had been scheduled and stored previously for future actions, such as a meeting that has not yet occurred.

Based on the above context information taxonomy, the state-of-the-art survey, the operator's expectations, typical scenario/application descriptions and the fundamental requirements for provisioning context-aware network services, Fig. 2.10 shows a representation of the context information taxonomy referred to and described above.

2.4.1.4 Representation of Information—Context Information Representation

Context modelling depends on the point of view of the context definition and scope. The model is a first approximation on how to structure, express and organize the context information as it is defined in [Dey00a] and [Dey01]. As mentioned before,

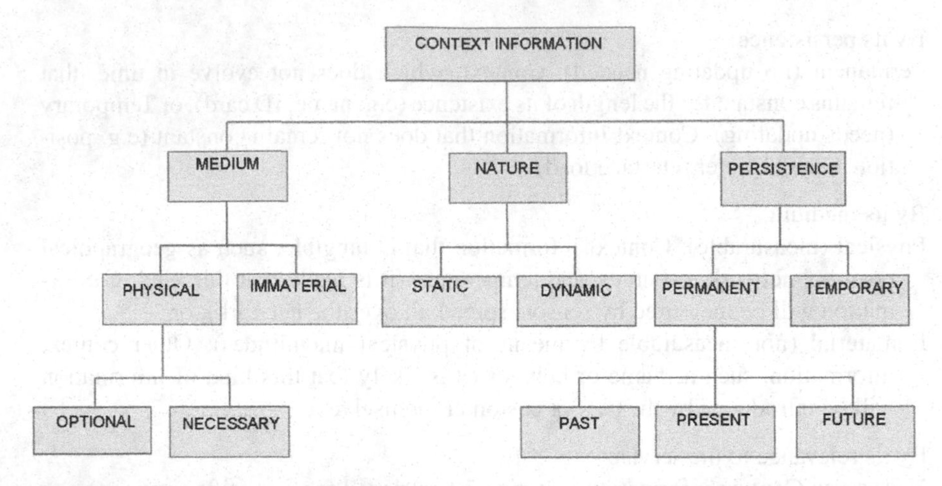

Fig. 2.10 Context information data model

the model is based on the concepts of entity and relationship and derived from the definition of entity in [Chen76].

An entity is composed of a set of intrinsic characteristics or attributes that define the entity itself, plus a set of relationships that are each instances of a standard set of relationship types. The concept of the local context of an entity can be defined as the information that characterizes the status of the entity. This status is made up of its attributes and its relationships. Moreover, the relationships that can exist between the different entities inside the model, as well as the entities themselves, can represent many different types of influences, dependencies, and so on, depending on the type of entities that these relationships connect.

With this type of model, one can construct a net of entities and relationships representing the world surrounding the activity of a context-aware service and thus the models can influence the development of the activity or service. This enables a scenario made up of many different types of information, and the influences or nexus that links one with the others. The local context enables the service to select and use context information from this scenario that is considered relevant in order to perform its task and deploy its service.

Figure 2.11 shows a simple, high-level example of modelling context by means of this entity-relationship technique [Serrano05]. In this example, two possible entities inside a hypothetical scenario are defined as a reference (an entity that represents a person and an entity that represents a printer device). The local context is defined as the sum of all of the specific attributes, plus the relationships established with other entities inside the model, of these two entities. Hence, this figure synthesizes the concept of local context of Person 1 or Printer 1, neither of which include entities such as Person 2 or Person 3. An entity becomes to be a part of local context in other entity when a relationship is defined, so Person 1 can be a part of local context of Printer 1 (right hand-side circle on the figure) and at the

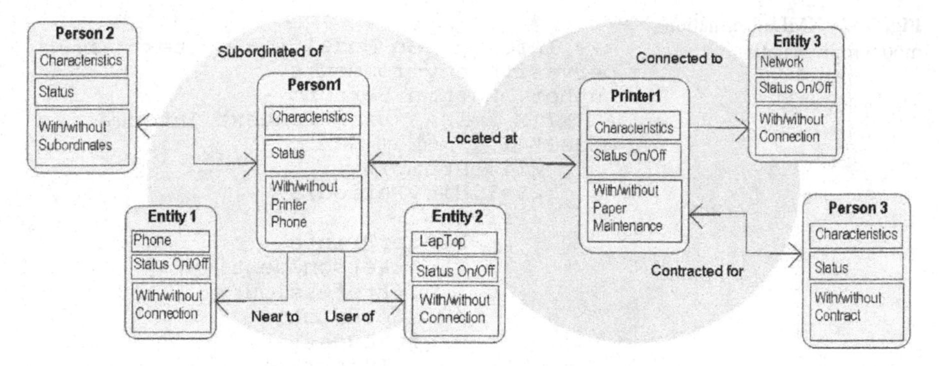

Fig. 2.11 High-level entity model with relationships example

same form Printer 1 can be part of local context of Person 1 (left hand-side circle on the figure).

2.4.1.5 Representation Tools for Information—Context Representation Tools

The tools that could be used to represent and implement the model, and the way to integrate this information model inside the general context-aware system architecture, need to be identified and tested, as they are potential tools for representing the context information. XML is a flexible and platform-independent tool that can be used in different stages of the information representation, which makes implementation consistent and much easier. The use of XML is increasing every day; however, it is by definition generic. Therefore, new languages that are based on XML have been developed that add application-specific features as part of the language definition. For example, to customize services, languages must have concepts that are related to the operational mechanisms of that service. It is in this context that XML Language has been successfully proposed and used to represent the context information models. XML has the following advantages:

XML is a mark-up language for documents containing structured information.

The use of XSD (XML schema definition) facilitates the validation of the documents created, even in a more basic but in some way also functional the use of DTD (document type definition) is also an alternative for validation. This validation can be implemented in a JAVA program, which can be the same used for creating and maintaining these XML schemas and/or documents.

Use XQuery, as a powerful search engine, to find specific context information inside the XML documents that contain all the information related to a specific entity. These queries can select whole documents or sub-trees that match conditions defined on document content and structure.

An example of using XML language for describing context information is shown in Fig. 2.12. This represents the context information model (CTXIM) as an example.

Fig. 2.12 XML information model representation

```
<xml version="1.0" encoding="UTF-8"?>
<-- Information Model for context in
pervasive environments.
Author: Martin Serrano -->
<CTXTIM xmlns:xsi ="Context Information
Model" xsi:noNam >
        <TimeStamp/>
        <ValidityPeriod/>
        <Person>
                <FingerPrints>
                        <PersonIdentity/>
                        <Profession/>
                </FingerPrints>
                <Preferences>
                        <Interface/>
                        <Service/>
                </Preferences>
        </Person>
        <Place>
                <Geographical/>
                <PhysicalSourround/>
        </Place>
        <Environment>
                <Time/>
                <Date/>
        </Environment>
        <Task/>
        <Object_Device>
                <Aplication/>
                <Network/>
                <Resource/>
        </Object_Device>
</CTXTIM>
```

Person, Place and Task entities are contained with specific descriptions as part of each entity.

2.4.1.6 Formalization of Context Information—Application of Ontologies

Context-awareness requires the following question to be solved: how can the context information be gathered and shared among the applications that use it? The answer to this question requires an extensible and expressive information model. The format to contain the information is part of the modeling process and the methodology to create the model. The most important challenge is to define the structure of the context information to collect, gather and store information. Context information can be used not just to model information in services, but also to manage the services provided. The model must be rich in semantic expressiveness and flexible enough to consider the variations of current status of the object being managed [McCarthy93]. The model should scale well with the network or the application domain.

A model considered reference in this book, about context modelling in pervasive computing environments, can be found and explained in [McCarthy97]. In this book the modelling result from this previous analysis and modelling activity formalizing information is referred as an excellent example. In other words and aligned with the objective of demonstrative facts pursued in this book, if the information models are expressive enough, pervasive systems can use that information to provide better management service operations. In order to formalize the information contained in the information model, ontologies appear to be a suitable alternative. However, this does not mean that other approaches are unsuitable for different applications. In this section, the idea to be discussed is that ontologies, in the field of management services, appear as a suitable alternative to solve the problem of formal modelling identified as providing the required semantics to augment the data contained in the information model in order to support service management operations. Ontology engineering can act as the mechanism for formalizing the information and provides the information with the semantic and format features, as it is the main scope in this section.

2.4.1.7 Distribution and Storage for Information Model—Context Model

The nature of context information is forcing systems developers to be able to provide computing facilities, such as context information data models, to anybody, anywhere, at any time. To provide these services, the infrastructures must be developed in a way that allows application programs running in the environment to efficiently locate the services. Currently, due to the numerous different applications that each use a different mix of technologies, each application tends to develop its own specific mechanisms to manage, store and process context information. In this section, an analysis based on generic approaches following their characteristics is briefly described.

Hierarchical Model

This organizes the data in a tree structure. This structure implies that a record can have repeating information, as each parent can have multiple children, but each child can only have one parent, and it collects all the instances of a specific record together as a record type. It is a clear example of dependency in higher hierarchical levels.

Network Model

This model organizes data as a lattice, where each element can have multiple parent and child records. This structure enables a more natural model for certain types of relationships, such as many-to-many relationships.

Relational Model

This model organizes data as a collection of predicates. The content is described by a set of relations, one per predicate variable, and forms a logic model such that all predicates are satisfied. The main difference between this organization and the above two is that it provides a declarative interface for querying the contents of the database.

Object/Relational Model

This model adds object-oriented concepts, such as classes and inheritance, to the model, and both the content and the query language directly support these object-oriented features. This model offers new object storage capabilities to the relational systems at the core of modern information systems that integrate management of traditional fielded data, complex objects such as time-series and geospatial data and diverse binary media such as audio, video, images and applets for instance. It also enables custom datatypes and methods to be defined.

Object-Oriented Model

This model adds database functionality to object programming languages. Information is represented as objects and extends object-oriented programming language with persistent data, concurrency control and other database features. A major benefit of this approach is the unification of the application and database development into a seamless language environment. This model is beneficial when objects are used to represent complex business objects that must be processed as atomic objects. As an example, object-oriented models extend the semantics of the C++, Smalltalk and Java object programming languages to provide full-featured database programming capability, while retaining native language compatibility.

Semi-structured Model

In this data model, the information that is normally associated with a schema is contained within the data, which is sometimes called "self-describing." In such a system, there is no clear separation between the data and the schema, and the degree to which it is structured depends on the application. In some forms of semi-structured models there is no separate schema, the schema itself can be converted if the data model is previously defined. In others models, the schema exist but only place loose constraints on the data. Semi-structured data is naturally modelled in terms of graphs.

Associative Model

The associative model uses two types of objects, entities and associations. Entities are things that have discrete, independent existence. Associations are things whose existence depends on one or more other things, such that if any of those things ceases to exist, then the thing itself ceases to exist or becomes meaningless.

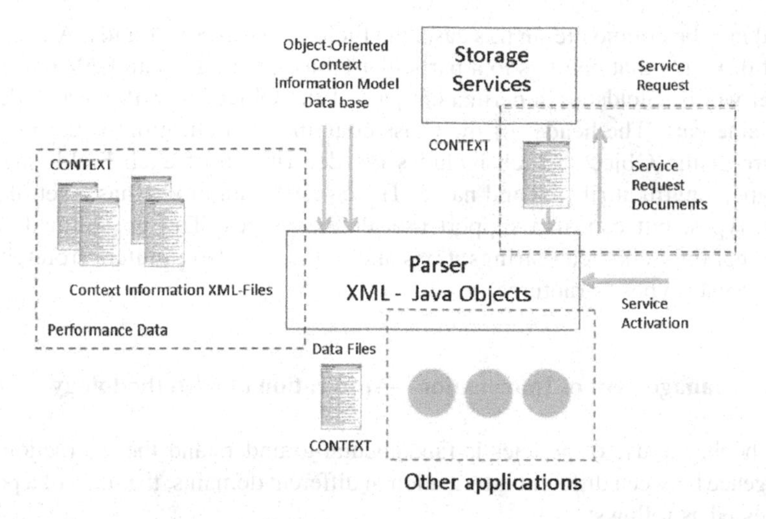

Fig. 2.13 Context information data model

Entity-Attribute-Value Model

The best way to understand the rationale of entity-attribute-value (EAV) design is to understand row modelling, of which EAV is a generalized form. Consider a supermarket database that must manage thousands of products and brands, many of which have a transitory existence. Here, it is intuitively obvious that product names should not be hard-coded as names of columns in tables. Instead, one stores product descriptions in a products table: purchases/sales of individual items are recorded in other tables as separate rows with a product ID referencing this table. Conceptually, an EAV design involves a single table with three columns, an entity, an attribute, and a value for the attribute. In EAV design, one row stores a single fact. In a conventional table that has one column per attribute, by contrast, one row stores a set of facts. EAV design is appropriate when the number of parameters that potentially apply to an entity is vastly more than those that actually apply to an individual entity.

In this section, different types of information databases have been studied. The context model is an alternative that seeks to combine the best features of different models; however, because of this combination, the context model is a challenge to implement. Thus, the possibility to use other data models is open. An object-oriented model, as shown in Fig. 2.13, has been implemented in Java, Extended explanation can be found in [Serrano06d]. The fundamental unit of information storage of the context model implemented in this book is, as described above, a Class, which contains Methods and describes Objects.

A consolidated context data model must combines features of some of the above models. It can be considered as a collection of object-oriented, network and semi-structured models. To create a more flexible model, the fundamental unit of information storage of the context model is a Class. A Class contains Methods and describes an Object. The Object contains Fields and Properties.

A Field may be composite; in this case, the Field contains Sub-Fields. A Property is a set of Fields that belongs to a particular Object, similar to an EAV database. In other words, Fields are a permanent part of the Object, and Properties define its variable part. The header of the Class contains the definition of the internal structure of the Object, which includes the description of each Field, such as their type, length, attributes and name. The context data model has a set of pre-defined types, but can also support user-defined types. The pre-defined types include not only character strings, texts and digits but also pointers (references) and aggregate types (structures).

2.4.1.8 Management of Information—Application of a Methodology

In this book, the use of policies is fundamental to understand the interaction and convergence between different information in different domains, the form of a policy is expressed as follows:

IF<conditions or events>THEN<x-actions>ELSE<y-actions>

Policies are used to manage the service logic at a higher level. A policy is used to define a choice in the behaviour of a pervasive service, and a pervasive service itself comprises a policy-based management system (PBMS). The adaptability of the PBM paradigm comes from its awareness of the operation environment, that is the context in which the management system and its components are being used.

A requirement for the management of information is the capability of the system offering service adaptation. When service adaptation occurs, there is an interesting change in the context. So a pervasive service management information model, which must be in reality an open, vendor-neutral approach to the challenge of technology change, represents an approach with a type of policy information model that also requires a language.

An interesting and extended proposal is to be implemented by using the model-driven architecture (MDA) initiative of the object management group (OMG) [OMG-MDA]. As requirement, a policy-based pervasive service specification language will also need to be used. The PBM methodology is to be implemented and demonstrated as a service composition platform and a service execution environment, both based on open source software and open standards. A set of service-centric network application programming interfaces (APIs) are developed by reusing existing user interfaces (UIs) to hide the heterogeneity of multiple types of end-user access networks and the core networks.

2.4.1.9 Implementation Tools

As far as PBM is concerned, this book do not try to develop new management techniques; rather, the research work is to apply existing PBM techniques to the network aspect of managing pervasive services in cloud environments. In this respect, it is wise to use the background of other standard-based information models and extend them to satisfy the services requirement needs (e.g. policy core information model (PCIM) [IETF-RFC3060] and [IETF-RFC3460] by the IETF, the core information model (CIM) by the DMTF [DMTF-DSP0005], and the Parlay policy information management (PPIM) APIs by the Parlay Group [Hull04]) between others.

In this book, specific emphasis on the policy-based descriptions for managing pervasive services is given, and a practically functioning pervasive service information model using ontologies to do so is pursued. The fact that there is not any reference implementation for the IETF PCIM [IETF-RFC3060], [IETF-RFC3460] or the DMTF CIM [DMTF-CIM] makes PBM hard to implement using these specifications. Hence, an step further in this section is to support the idea to enhance and control the full service life cycle by means of policies. In addition, this approach takes into account the variation in context information, and relates those variations to changes in the service operation and performance inspired from autonomic management principles and its application in cloud environments.

2.4.2 The User Requirements

This section describes the envisaged requirements for services from each of the parties involved in the service value-chain. The benefits can be obtained for each of these stakeholders with the introduction of context information models that represent the effective functionality of pervasive services in a vendor- and technology-neutral format as summarized. This section refers to work undertaken by the author as collaboration research activities into the EU IST-CONTEXT project and the EMANICS Research Network; the public documentation that defines the state of the art of context-aware services in the pervasive computing knowledge area can be found in the Web sites [IST-CONTEXT] and [IST-EMANICS]. Here is presented just a resume of those main user requirements since the scope that they are necessaries to any pervasive service being managed by policy-based managed systems.

2.4.2.1 End-User Requirements

- Cheap end-user devices that provide compelling new functionality.
- Device manufacturer independence.
- Selection from an extensive range of new services.
- Personalized/smarter services

2.4.2.2 Service Providers Requirements

- Personalized services allowing customization of respective service portfolios.
- Significant technological and operational savings as a result of standardization of context information models.
- Significant added revenue from context-aware services.

2.4.2.3 Content Providers Requirements

- Exploitation of context information resulting from common platforms using standardized context information models.
- Flexible entering of partnerships facilitated through standardized context information models.
- The freedom to use any particular service provider for the advertising and providing of content.

2.4.2.4 Network Providers Requirements

- Significant technological and operational savings by using standardized context information models.
- Common equipment requirements and support for CAS based on standard context information models.
- New business opportunities that result from context-aware services.

2.4.3 The Technology Requirements

The tendency in modern ITC systems is that all control and management operations are automated. This can be effectively and efficiently driven by using context information to adapt, modify and change the services operation and management offered by their organizations. The context-awareness property necessarily implies the definition of a variety of information required to operate services in next-generation networks.

This section relates and describes the technological requirements to support such ITC context-aware features and identifies the properties of pervasive applications for supporting service management operations in cloud environments. The pervasive properties describe how management systems use context information for cross-layer environments (interoperability) in NGNs to facilitate information interoperability between different service stake holders (federation). The analysis about using information following such technological requirements is a task of this book to exemplify and help as reference to satisfy one or various cloud service management aspects.

The priority of this section is the support of multiple and diverse services running in NGNs. Such scenario(s) are typified by complex and distributed applications, which in terms of implementation and resource deployment, represent a high management cost due to the technology-specific dependencies that are used by each different application. This in turn makes integration very difficult and complex.

Before providing a definition of technology requirements, Fig. 2.12 shows the information requirements for supporting services. This shows the perspective of technology requirements, their relationships and the level of influence between the

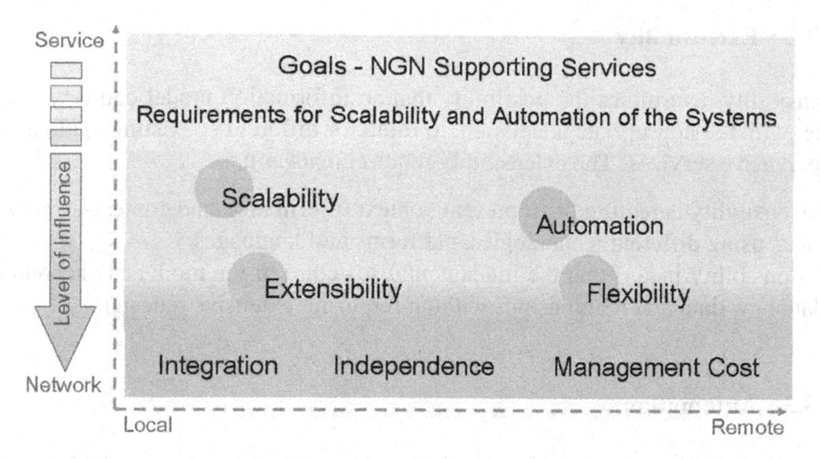

Fig. 2.14 NGN supporting services, relationships and influence

properties that enable the advantages of using context information to be realized by management systems.

Figure 2.14 depicts the level of influence of NGN requirements supporting information services, and how those are related to the service and network views, respectively. On the left hand-side, the arrow shows how the information collected from top-service layers can be used to modify network conditions affecting the different technology requirements of the information management systems.

This section surveys the envisaged requirements for NGNs supporting information services [Serrano06b]. Those requirements are related to management systems supporting services and using integrated information. The requirements are related to the scalability and automation of the systems. These properties will produce increased extensibility and flexibility, as well as produce attendant network and service management cost reductions. They also provide implementation advantages for services as follows.

2.4.3.1 Scalability

Information systems solutions need to scale with the number of users, services and complexity in acquiring and filtering the context information used by the pervasive application. Information needs to be transparently distributed among and along the applications within heterogeneous service environments. The scalability requirements are:

- High levels of scalability due to the inherent necessity to represent diverse types of information within multiple services.
- The necessity to extend the definition and representation of information to other platforms supporting different services.
- Solutions that scale according to the number of users and services in acquiring and filtering information.

2.4.3.2 Extensibility

Extensibility determines the possibility that an information model could be applicable, with the appropriate adaptation, to future or different coexisting applications for pervasive services. The extensibility requirements are:

- Extensibility is required to represent context information and context-aware services using different technologies, platforms and languages.
- Extensibility means that the fundamental structure of the model can accommodate new data and relationships without requiring extensive redesign.

2.4.3.3 Automation

Automating services aims to provide tools and mechanisms for automatic service creation, operation and management. This enables the system to reduce or suppress the need for any human intervention. Information models, supported by updating data mechanisms, can significantly contribute to reach this goal. This implies:

- Mechanisms that use information for automatic service creation, operation and management of the information and services must be supported by the information model.
- Reduce as much as possible the need of any human intervention to manage services.
- Software solutions for supporting self-* operations in terms of diverse user-centred services.

2.4.3.4 Flexibility

Flexibility is the characteristic or capability of the systems for adapting the service creation, deployment, customization and operation to market and user needs. The information model can contribute to this service characteristic by providing the necessary abstractions of the underlying network services and network infrastructure to the entities and stakeholders involved in the service lifecycle. The requirements of this feature are:

- The capability for adapting the service lifecycle management operations to market and user needs must be able to be represented in the information model.
- The necessary abstraction mechanisms to represent the underlying network resources and services must be able to be represented in the information model.
- Middleware solutions that enable business goals to determine information-specific network services and resources must be able to easily use data in the information model.

2.4.3.5 Integration

As a result of new services being defined by independent providers, rather than by the segmentation of specific markets, fragmented approaches using different types of services are offered by different providers using different implementation techniques in different infrastructures. The danger is that the interaction desired by end users is no longer possible due to this fragmentation. A single extensible information model is required that can harmonize the different application-specific services and context information of the different providers. The requirements of this feature are:

- Adaptation of the information systems for providing the information with a format according to the information model. As a result of adaptation, new technologies can support information services from different providers with different implementation mechanisms.
- An extensible information model to enable integration with systems that are designed in a stovepipe fashion.

2.4.3.6 Independence

It is well known that vendor independence is a feature desired for all systems. Hence, the requirements for this feature are:

- Service providers need independence from equipment manufacturers, which promotes the interoperability of the information that they supply for supporting services.
- The information model must be able to model functionality from different vendors that operate on different platforms and use different technologies.

2.4.3.7 Management Cost

Telecommunication services are only viable if they do not incur recurring capital and operational expenditures. The additional traffic and data heterogeneity caused by application-specific information has, up to now, increased operational expenditure, since it requires custom middleware or mediation software to be built to integrate and harmonize disparate information from different applications. Hence, the requirements are:

- Reduce management cost by lowering the number of skilled resources for managing heterogeneous pieces of information.
- Reduction of side effects when implementing information models that use and integrate heterogeneous pieces of information.

2.4.4 *Cloud Computing Challenges and Trends*

As premises in cloud computing, the reduction of payment cost for services and the revenue benefit for investment in proprietary infrastructure are the key factors to believe cloud is the solution to many of the under usage problems and waste in technology the IT sector is facing up. Particular interest for cloud computing services and the use of virtual infrastructure supporting such services is also the result of business model which cloud computing offers, where bigger revenue and more efficient exploitation are envisaged [IBM08]. Likewise, there exist a particular interest from the industry sector, where most of the implementations are taking place, for developing more management tools and solutions in the cloud, and in this way the pioneers are offered full control of the services and the cloud infrastructures. On the other hand, far away from revenue benefits, academic communities point towards finding solutions for more powerful in terms of computing processing and at the same time more efficient to reduce enormous headaches when different technologies need to be interactive to exchange a minimum part of information. Thus, generally problems on manageability, control of the cloud and many other research challenges are being investigated.

Cloud computing management is a complex task [Rochwerger09], for example clouds must support appropriate levels of tailored service performance to large groups of diverse users. A sector of services, named private clouds, coexist with and are provisioned through a bigger public cloud, where the services associated to those private clouds are accessed through (virtualized) wide area networks. In this scenario, management systems are essential for the provisioning and access of resources, and where such systems must be able to address fundamental issues related to scalability and reliability issues which are inherent when integrating diverse cloud computing systems.

2.4.4.1 Monitoring in the Cloud

Monitoring in a cloud is essential for automatic or autonomous adaptation to current load, as well as to provide feedback on SLA fulfilment. Scalability and security are essential for cloud monitoring. Without solving the problems of scalability and security, tools and technologies are almost impossible to be deployed in a cloud environment. Moreover, cloud monitoring will also require application-level information to be monitored in addition to the system usage information which current tools can provide [Shao10].

There are several related works in this area. OpenDS [OPENDS] introduces system monitoring for distributed environments and mainly focuses on fault tolerant aspects; NWS [Wolski99] also provides a distributed framework for monitoring and has the ability of forecasting performance changes; When compared to DSMon and NWS, another system resources monitoring tool called DRmonitoring [Domingues03] requires less resources to run and supports multiple platforms (Linux and Windows).

However, up to date of the publication of this book, the DRmonitoring tool lacks scalability and fails to address security concerns. From the industrial viewpoint, HP Open View [HPOPENVIEW] and IBM Tivoli [IBMTIVOLISIC] have been developed to ease system monitoring and are primarily targeting thc enterprise application environment. Although, the commercial products are relatively limited in portability across different operating systems, they are usually highly integrated with vendor-specific applications, subject to new versions released after the edition of this book.

In the cloud environment, heterogeneity is one of the fundamental requirements for monitoring tools. Therefore, the industrial tools are unsuitable for more general purpose monitoring of the cloud. GoogleApp engine [GOOGLEAPP] and Hyperic [HYPERIC] both provide monitoring tools for system status such as CPU, memory and processes resource allocations. Such system usage data can be useful for general purpose cloud monitoring, but they may not be sufficient enough for an application-level manager to make appropriate decisions.

To address the limitations of existing tools, a new monitoring tool named runtime correlation engine (RTCE) [Holub09] has been developed at the UCD Performance Engineering Laboratory jointly with IBM Software Verification Test teams. RTCE takes the important concerns of heterogeneity, high performance, low overhead and need for reasonable scalability. In the near future, RTCE can be essentially improved with greater scalability based on several proposed architectures [Wang10]. RTCE will be primarily focusing on stream data correlation and provide flexible data results for other system components to consume. The output data produced should be generic and scalable, so that any type of component can easily adopt the content of the data. In the case of changing signatures on the output, the existing applications should still be able to consume the new output data without any code changes.

2.4.4.2 Non-expert Monitoring of Cloud Services

The need for end users to become involved in cloud infrastructure, service monitoring and management is driven by two key features of cloud computing. As the number of virtual resource "instances" and individual features of resources and services continue to grow to provide flexibility, it will become increasingly unrealistic to expect the cloud provider to manage resources and services for end users.

In addition, as users will pay for resources in a very fine-grained manner, end users may want to monitor their own resources so they can decide when to request more/less resources and ensure they are getting an optimal value for the resources they pay for. A model where user management and monitoring preferences and requirements are mapped to the underlying resources and services in a constrained yet extensible manner, thereby giving users to control over the resources they used to pay.

Harmonization of monitoring data requires mechanisms for mapping of the large volume of low-level data produced by resource-level and service-level monitoring systems into semantically meaningful information and knowledge that may be discretely consumed by non-experts. Resource-level and service-level experts are generally very familiar with how the constituent parts of the system can be managed, and are particularly aware of the end-to-end operating constraints of those constituent parts. Encoding this expertise in a manner that can be utilized by other stakeholders would enable stakeholders of other systems to monitor other parts of the system and relate their operation to their own system. It also enables common knowledge to be shared and reused.

Supporting personalized visualization of harmonized monitoring data for non-expert end users is viewed as a key to include non-technical "prosumers" in the end-to-end service quality of experience-based control loop [Hampson07]. The relevance of particular monitoring information is dependent on the user-level, application-level, and possibly environmental context, and hence must be properly contextualized [Novak07] in a way that the data consumer appreciates its relevance.

This has led to the popularity of personalized "dashboard" type monitoring applications for visualizing key performance indicators of managed systems [Palpanas07]. A key factor of this approach is to support cloud computing customers in their strategic decisions to invest in more targeted resources, and satisfy themselves that they are getting sufficient value from their investment in cloud computing.

2.4.4.3 Federation–Cloud Interconnection

Federation in the cloud would imply a requirement where user's applications or services shall still be able to execute across a federation of resources stemming from different cloud providers. It also refers to the ability for different cloud providers to scale their service offerings and to share capabilities to combine efforts and provide a better quality of service for their customers. While the technological aspects required to support cloud federation is an ongoing research domain, there has been little work to support the holistic end-to-end monitoring and management of federated cloud services and resources. This approach requires users and multiple providers to both delegate, share and consume each other resources in a peer-to-peer manner in a secure, managed, monitored and auditable fashion, with a particular focus on interoperability between management and resource description approaches.

Federation in the cloud has the potential to enable new services across a growing range of communication and computing infrastructures to be defined by different providers (or other administrative entities), such that the result enables a transparent service to be available to the end user. Here, "transparent" means that the end user is not aware that the service he or she is using is actually provided by multiple providers. Hence, an important connotation for Federation is that services that a Federation provides should appear to be produced by a single entity. However, this raises two orthogonal challenges: how to support transparent and seamless interworking/sharing or resources and services for users, and how support operators to securely monitor, manage and share each others' heterogeneous resources to achieve this.

Federation represents an approach for a solution supporting the increasingly important requirement to orchestrate multiple vendors, operators and end-user interactions [Bakker99], [Serrano10], and it is clear in the applicability of this concept in the cloud computing area.

Cloud computing offers an end-user perspective where the use of one or any other infrastructure is transparent, in the best case the infrastructure is ignored by the cloud user [Allee03]. However, from the cloud operator perspective, there are heterogeneous shared network devices as part of diverse infrastructures that must be self-coordinated for offering distributed management or alternatively centrally managed in order to provide the services for which they have been configured. Furthermore, there must be support to facilitate composition of new services which requires a total overview of available resources [Kobielus06].

In such a federated system, the number of conflicts or problems that may arise when using diverse information referring to the same service or individuals with the objective of providing an end-to-end service across federated resources must be analyzed by methodologies that can detect conflicts. In this sense, semantic annotation and semantic interoperability tools appears as tentative approach solution and that currently is being investigated.

2.4.4.4 Elasticity—Management of the Cloud

The need to control multiple computers running applications and likewise the interaction of multiple service providers supporting a common service exacerbates the challenge of finding management alternatives for orchestrating between the different cloud-based systems and services. Even though having full control of the management operations when a service is being executed is necessary, distributing this decision control is still an open issue. In cloud computing, a management system supporting such complex management operations must address the complex problem of coordinating multiple running applications' management operations, while prioritizing tasks for service interoperability between different cloud systems.

An emerging alternative to solve cloud computing decision control, from a management perspective, is the use of formal languages as a tool for information exchange between the diverse data and information systems participating in cloud service provisioning. These formal languages rely on an inference plane [Strassner07b], [Serrano09]. By using semantic decision support and enriched monitoring information management, decision support is enabled and facilitated. As a result of using semantics, a more complete control of service management operations can be offered, hence a more integrated management, which responds to business objectives. This semantically enabled decision support gives better control in the management of resources, devices, networks, systems and services, thereby promoting the management of the cloud with formal information models [Blumenthal01].

This section addressed the need to manage the cloud when policies are being used as the mechanism to represent and contain description logic (DL) to operate operational rules. For example, the SWRL language [Bijan06], [Mei06] can be used to formalize a policy language to build up a collection of model representations with the necessary semantic richness and formalisms to represent and integrate the heterogeneous information present in cloud management operations. This approach relies on the fact that high level infrastructure representations do not use resources when they are not being required to support or deploy services [Neiger06], [VMWARE]. Thus, with high-level instructions, the cloud infrastructure can be managed in a more dynamic and optimal way.

2.4.4.5 Support for Configuration and Re-configuration of the Cloud

Several cloud usage patterns can be identified based on bandwidth, storage, and server instances over time [Barr10]. Constant usage over time is typical for internal applications with small variations in usage. Cyclic internal loads are typical for batch and data processing of internal data. Highly predictable cyclic external loads are characteristic of Web servers such as news, sports, whereas spiked external loads are seen on Web pages with suddenly popular content (cf. "slashdotted"). Spiked internal loads are characteristic of internal one-time data processing and analysis, while steady growth over time is seen on startup Web pages.

The cloud paradigm enables applications to scale-up and scale-down on demand, and to more easily adapt to the usage patterns as outlined above. Depending on a number or type of requests, the application can change its configuration to satisfy given service criteria and at the same time optimize resource utilization and reduce the costs. Similarly clients—which can run on a cloud as well—can re-configure themselves based on application availability and service levels required. On-demand scalability and scalability prediction of a service by computing a performance model of the architecture as a composition of performance models of individual components are also features to be considered when designing cloud solution. Exact component's performance modelling is very difficult to achieve since it depends on a various variables such as available memory, CPU, system bus speed and caches.

2.5 Conclusions

In this chapter…

The requirements of information and pervasive service, based on both NGN demands on context information and the demands of context-awareness according to a set of pervasive service requirements, have been studied and discussed.

The use of ontologies as the formal mechanism to integrate context information for managing service operations is the most adequate. Likewise ontologies is adequate for other management-related activities as policy composition, resources management definition and control and service application representation. In summary the most suitable alternative for supporting interoperability in cross-layer systems. The use of object-oriented models and the policy-based paradigm, both with ontology-based representation, act as a premise in this book. The combination of technologies seems to be a feasible alternative for meeting the convergence requirements of information systems making use of services in NGNs at extensively in cloud environments.

The discussed approaches for ontology-based modelling of context information in pervasive applications cover the information requirements for modelling and structuring of the information in a formal, extensible and efficient manner to collect, distribute and store information. However, current approaches ignore the importance that context information has when it is integrated with management operations. Therefore, a valuable contribution of this chapter is the easy understanding about capabilities and use of information to define and represent management operations.

The main benefit of using policies is to manage the complexity of the management of resources and services in an extensible, scalable and flexible manner. The simplification, which is obtained by providing higher-level abstractions as a result of using ontologies, is also an advantage. Policies are created to help automate the whole system, allowing the adaptation of the services offered according to changing context. This is a crucial goal of pervasive computing systems. It is important to note that this is a unique proposal in this area.

Cloud computing challenges and trends has been studied and discussed in the framework of infrastructures to support cloud services research efforts to provide a clear and valuable state-of-the-art survey, and analysis of cloud trends and challenges has been introduced and discussed. Thus, the use of knowledge platforms that represent and integrate information in various management operations and potentially cloud environments is perfectly justified.

Chapter 3
Using Ontology Engineering for Modelling and Managing Services

3.1 Introduction

This chapter reviews basic concepts about ontology engineering, with the objective of providing a better understanding of building semantic frameworks using ontologies in the area of telecommunications and particularly managing communication services. This background enables the reader to better understand how ontologies and ontology engineering can be applied in communications, since both of these have only recently been applied to this field. Ontologies are used to represent knowledge, and ontology engineering is made up of a set of formal mechanisms that manage and manipulate the knowledge about the subject domain in a formal way endowing semantic riches to the information.

Ontology engineering has been proposed as a mechanism to formalize knowledge. This chapter presents those basic concepts referenced in this chapter as inherent features for supporting network and services management. Alike defines the lexical basic conventions for the process of creating ontologies helping the reader to understand how ontology engineering can augment the knowledge and support decision making of current management systems.

The organization of this chapter is as follows. Section 3.2 introduces a general scope for introducing the basic elements for building up ontologies and how ontologies are structured. It provides ontology engineering definitions for concepts and relationships, and describes the representation tools and functions. Instances as well as axioms that are being used to build ontologies within this chapter are also explained.

Section 3.3 provides general understandings about pervasive services and semantics in a form of introductory basic definitions for concepts related, and describes its usage implications in different areas to establish this concept in the state of the art.

Section 3.4 reviews various semantic operations that can be done with ontologies. The basic ontology operations, as tools to support the management systems, can execute semantic control of context information. This section briefly describes those operations that management systems can do using context information when

J.M. Serrano Orozco, *Applied Ontology Engineering in Cloud Services, Networks and Management Systems*, DOI 10.1007/978-1-4614-2236-5_3,

ontologies are used for the formal representation and modelling of knowledge in cloud systems.

Section 3.5 presents a review of two different types of functions that computing systems can perform with ontologies. The first group consists of ontology mapping, merging and reasoning tools, for defining and even combining multiple ontologies. The second group uses ontologies as development tools, for creating, editing and managing concepts that can be queried using one or more inference engines.

3.2 Ontology Structures: Elements and Links

The stage of the research presented in this section has been achieved by seeking how to satisfy the requirements dictated by pervasive services regarding information interoperability. Strang and Linnhoff-Popien [Strang04] have described and present a classification, about how much they satisfy or contribute to semantic requirements enrichment. This classification is done in terms of information modelling capabilities and particularly context information models are used for supporting services interoperability. The idea that ontologies offer more capabilities for satisfying different information requirements in terms of semantic richness as it is discussed in [Strang03b] is supported. While ontologies do have some shortcomings, mainly in the consumption of more computing resources, in the final analysis, advantages overcome the drawbacks or restrictions.

The section about ontology structures acts as an introduction to using ontologies as the formal mechanism for integrating and increasing the semantics of facts represented in information models, in this book such facts are particularly referred to management of services and networks.

A state of the art on ontology categorization, as well as a hierarchical description for applying ontologies for integration of concepts or information models, has been formerly presented in [López03a] and [López03c]. In this chapter, the research efforts are addressed towards the correct application of ontologies for representing knowledge into pervasive applications for service and network management as well as services life cycle management operations control and the functional architecture supporting such applications. However, to date, no other approaches have deeply analyzed the meaning and significance of management data using ontology-based context information and their implications when using ontology in other engineering disciplines.

As have been defined in [Gruber93b] and [Guarino95], ontologies have been used to represent information that needs to be converted into knowledge. In the cognitive conversion process, the information is formalized as a set of components that represent knowledge about the subject domain in a general and non-specific manner [Gómez99]. This representation then leads to a formal coding approach. In this section, these components are described. Ontologies can be used to support service and network management goals and also management operations and process. Ontologies can be extended to provide application- and domain-specific ontologies that augment information and data models to meet the needs of next-generation network (NGN) and service management.

Ontologies provide the necessary formal features to define a syntax that captures and translates data into ontological concepts; otherwise, syntax can only be weakly matched (i.e. using patterns). In the domain of communication systems, the syntax must be formalized in a functional way with the objective of supporting NGN management operations [both operational support systems (OSS) and their associated business support systems (BSS)] and as a consequence of supporting the new pervasive services creation, deployment and management.

Ontologies are formally extensible. Systems can take advantage of this extensibility and semantic interoperability for supporting the management system. In addition, when integrating context information into pervasive services, context will help specify management services more completely, as well as formally adjust and manage resources and services to changes in autonomic environments.

This chapter pays special attention for focusing pervasive service applications in the framework of autonomic communications. Thus, the use of ontologies offers significant benefits in terms of representing semantic knowledge, and reasoning about that information is being managed. This helps to promote the information interoperability behind the integrated management.

3.2.1 Concepts

Concepts are the abstract ideas that represent entities, behaviour and ideas that describe a particular managed domain. Concepts can represent material entities such as things, actions and objects, or any element whose concepts and/or behaviour needs to be expressed by defining its features, properties and relationships with other concepts. Such concepts can be represented and formalized as object classes. The classes are used and managed by computing systems for performing operations or simply for sharing information.

3.2.2 Representation

The representation is a formal or informal way to understand and situate the idea in reference to certain properties or features in the domain where the idea is created. The representation can be created using formal tools or mechanisms for depicting the ideas or concepts, or the representation can be informal, such as using a simple graph or a set of symbols depicting the ideas of the concept.

3.2.3 *Relationships*

Relationships represent the interaction between the concepts of a domain. These include structural agreements (e.g. subclass-of and connected-to in the field of pervasive computing) as well as semantic descriptions (e.g. synonyms, antonyms and is-similar-to) that can be used to express how a concept interacts with other concepts in the managed domain.

3.2.4 *Functions*

Functions are a specific type of relationship in which an element is identified as the result of a process or activity. Functions are not necessarily limited to mathematical functions, and can, for example include logic functions for defining relationships in the form of conditionals or aggregations.

3.2.5 *Instances*

Instances are used for creating specific objects already defined by a concept and can represent different objects of the same class (e.g. person 1 instance-of and person 2 instance-of). Instances enable objects that have the same properties, but are used to represent different concepts, to be realized, and can be described as sub-components of concepts that have already been modelled.

3.2.6 *Axioms*

Axioms are the logic rules that the ontology follows. Axioms are theorems that contain the logic descriptions that the elements of the ontology must fulfil. The axioms act as the semantic connectors between the concepts integrating the ontology, and they support the logic operations that create a dynamic interaction between the concepts. In pervasive computing, the axioms act as conditions for linking the concepts and create the functions between concepts in an ontology.

3.3 Semantics and Pervasive Services

Pervasive computing and semantic Web is acquiring more and more importance as a result of the necessity for integrating context information in service provisioning, applications deployment and network management. In today's systems, making context information available in an application-independent manner is a necessity.

The expansion of the semantic web and the consolidation of pervasive computing systems have as result an increasing demand on the generation of standards for service-oriented architectures (SOA) and also on Web Services. The target in current pervasive systems is the use of the environment itself, in a form that the information can be easily accessible for the support of services, systems and networks. In those scenarios, systems need to be prepared to use diverse information from multiple information models, and most importantly, model the interaction between the information contained in business models and network models.

Service management platforms must be able to support the dynamic integration of context information, and thus take advantage of changing context information for controlling and managing service and network management operations in service provisioning operations. To achieve this kind of service management, flexible architectures are necessary that are based on the knowledge and construction of relational links between principal concepts, using relationships to build extensible semantic planes founded on a formal definition of the information.

The creation of a semantic plane following SOA demands the combination of multiple and diverse technologies, but principally agents. Programmable networks and distributed systems provide the background necessary for implementing pervasive services in NGNs.

The state of the art, presented in this section, is concentrated on management failures. Diverse are the reasons makes a management system inefficient, part of them are listed as trends in service management and computing and referred in state-of-the-art sections.

In the framework of this chapter, and as main integrated management objective, three different types of failures already identified, acting as guidance: (1) technological failures as a result of hardware problems, (2) hardware limitations creating management errors as a result of limited capability for processing information (e.g. overload of the systems when multiple and diverse systems are being managed as a result of different data models) and (3) middleware limitations for exchanging information between systems as a result of diverse technologies using different information and data models. Currently, management systems are able to detect those failures and follow pre-defined procedures to re-establish the communications infrastructure.

However, a most important type of failure exists that is related to content and semantic issues. These failures occur when management systems operate with wrong information, when data is changed erroneously after a translation or conversion process, or when data is misinterpreted or not fully understood. Such problems are still largely unsolved. However, the introduction of ontology engineering is seemed as a tool or mechanism to help solve these problems.

In Fig. 3.1, the three areas of concern (pervasive management, context data and communications systems domain) are identified. The domain interactions clarify the actions that this section addresses. Pervasive management provides the interfaces and mechanisms to users and system applications, enabling them to utilize services as a result of variations in the context information. The communications systems domain refers to software and hardware components needed in the operation of management services, and for their deployment and execution.

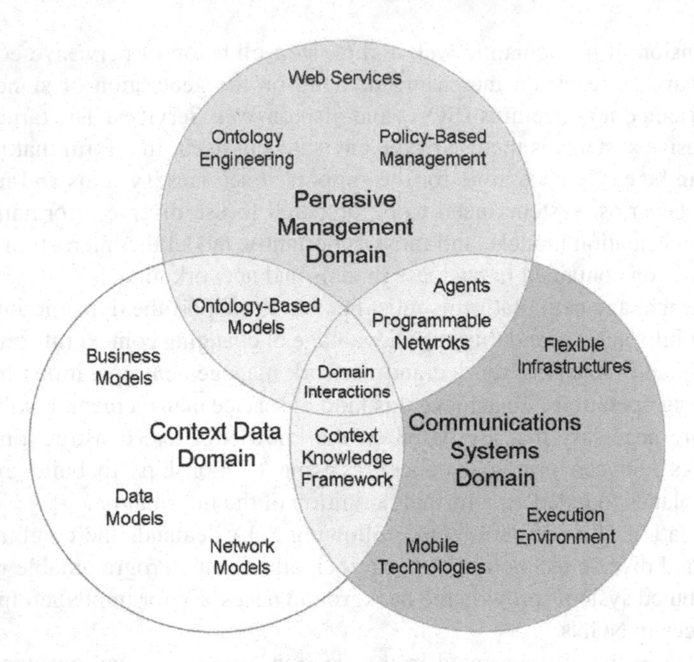

Fig. 3.1 Semantic domain interactions in pervasive computing

The context data domain provides all the formal data mechanisms, business-oriented as well as network-oriented, to represent and handle the information related to users and networks that is used in management operations to support pervasive applications.

3.3.1 *Understanding Context-Awareness and Its Usage Implications*

As our computing environment evolves, the definition of context-awareness evolves as well. The definition from Dey et al. [Dey00a] more or less reflects this change: "A system is context-aware if it uses context to provide relevant information and/or services to the user, where relevancy depends on the user's task". Dey et al. have chosen a more general definition of context-aware computing.

The definitions in the more specific "adapting to context" category require that an application's behaviour be modified for it to be considered as context-aware. When tried to apply these definitions to establish context-aware applications, it was found that it did not fit. For example, an application that simply displays the context of the user's environment to the user is not modifying its behaviour, but it is classified as being context-aware. If a less general definition is used, these applications would not be classified as context-aware. Therefore, a more general definition that does not exclude existing context-aware applications was proposed. "Context-awareness refers to the properties of a system that make it aware of its users' state and network surroundings and help it to adapt its behaviour conveniently".

The notion of context-awareness can make a system more efficient and less intrusive, an important characteristic of pervasive computing systems (e.g. in a given service where the user needs to choose between various options, context-awareness can reduce the number of decisions required to choose the correct option. Under this simple scenario, the advantages of requiring some (minimal) information outweigh the disadvantages of being intrusive). A context-aware system can thus be seen as an assistant to help the user make decisions in a proactive fashion. It can also be used to anticipate user needs [Korke-Aho00]. The objective in pervasive computing is to construct context-aware systems that can emulate such intelligent decision making, and only interrupt the user in an emergency or a situation that has not been encountered before.

Context-awareness enables a new class of service applications in the research and development area, known as pervasive computing or ambient intelligence. There are many definitions that are used; the following provides the scope and terminology of the key concepts in this area which are used in this chapter.

As defined before, context represents any information that can be used to characterize the situation of an entity [Dey00a]. An entity can be defined as the person, place or object that is considered relevant to the interaction between a user and an application, including the user and application themselves. This adaptation to the definition reflects the inherent heterogeneity of information that makes up context. For example, context may refer to aspects of the physical world as well as conditions and activities in the virtual world.

The information referred to as the status, performance or situation of the network is of great importance, and it is known as contextual network information. Many different initiatives have been developed for defining, modelling and using network contextual information with or without end-user context information, as described in [Gellersen00], [Dey00b], [Ducatel01], and in [Dey97], [Dey99], [Nakamura00]. While each has its own scope, all of them consider context as relevant and important to provide the services or the correct functionality of the application.

However, none of these examples use the information to modify the service. In this chapter, context-awareness in network services can be understood as the ability to use context information to provide a better service, taking into account the information to build decisions and even react to changes of the network, user and service context. In this manner, it is expected that services act in a concerted mode.

Context-awareness cannot be achieved without an adequate methodology and a suitable infrastructure. Hence, one of the goals of this chapter is to provide a solution in the form of a semantic plane that is architected to make the management of pervasive services feasible, efficient and flexible by basing that management on a context information model. Web services, policy-based management (PBM) and programmable network technologies are integrated for this purpose.

3.3.1.1 State of the Art in Context-Awareness

Context has an important role in the design of systems and/or applications that are context-aware, since any change in context will change the functionality provided. In the same manner that a gesture or a word can have different meanings depending on the context or the situation in which they are expressed, a context-aware service can also act differently. Since context can be made up of logical or physical characteristics, and expressed as different levels of abstractions, it can be perceived as either a physical or virtual effect that alters the functionality and/or performance of context-aware applications.

Context information must be collected and properly managed to make pervasive services a reality; due to its inherent dynamic nature, this poses extreme demands on the network and the service layers that most current technologies are unable to fulfil. Work in this field has been concentrated on creating, collecting and deploying context sensitive information as described in [Dey98], [Schmidt01] to the user. Context information is described as the knowledge about the user's and/or device's state, including its surroundings, location and, to a lesser extent, situation.

Previous context-aware network services have been mostly focused on mobile users that may see their context changing due to changes in location (also termed as location-based services (LBS)). For example, [Finkelstein01] describes efforts to design, specify and implement an integrated platform that will cater to a full range of issues concerning the provisioning of LBS. This solution is made up of a kernel and some support components that are in charge of locating the user and making services accessible. Within the scope of this solution, there is a service creation environment that will enable the specification, creation and deployment of such services within the premises of service operators; however, those solutions do not consider the control of operations as a crucial activity.

Other approaches have been developed that aim to design and implement technology that is personalized to the users and sensitive to their physical situation [Long96]. The idea is to develop a new system architecture, which will enable ambient information services to be delivered to mobile citizens [Schilit95]. For user's location purposes, a set of sensors is deployed. Moreover, these sensors will detect and send contextual information about the surroundings of the mobile users. These sensors can be networked and integrated within existing computers and wireless network infrastructures. In these approaches, the importance of using context information for controlling the services is unfortunately not emphasized and useless for control management operations.

There are research efforts that relate context information and use it for other purposes than only representing information [Schilit94b]. Other examples of research activities for end-devices are the composite capabilities/preference profiles framework [CCPP] and applications with agent and profile specifications [Salber99], [Gribble00], [Gruia02]. Example research also exists in the IETF for networks, including the open pluggable edge services [OPES], content distribution interworking [IETF-CDI] and Web intermediaries [IETF-WI] working groups, which develop frameworks and recommendations for network communications, especially for content peering and for adaptation purposes by using and processing context information to change the performance of the applications.

One implementation, provided in existing projects, follows the client–server principle, incorporating a server with numerous clients to provision and deliver multimedia information concerning the user's location and orientation. This system is able to determine the client's position in the overall network, to manage personal data of each user and each task by associating geographic information system (GIS) data to multimedia objects and to provide ubiquitous services; another important characteristic of this project is the design and implementation of a user's new generation mobile terminal device [LOVEUS].

However, location is not the only context information that is being used. For example, some of these initiatives have focused on designing a toolkit that provides general and modular solutions for mobile applications by adapting the content to the device that will use it [Gellersen00], [Samann03]. With that purpose, they propose a software toolkit that will be hosted on the user's terminal and will adapt the format of the content, usually multimedia, to specific capabilities of the terminal device.

Finally, it is important to mention initiatives oriented to locate the user by establishing a network of beacons located in the building where the user can move around. From these beacons, the user's device will be able to know its location [UAPS]. There are some other projects, in which the main goal is to introduce ubiquitous computing in everyday environments and objects [Fritz99], [Hong01].

Other attempts in the use of programmable network technologies for context-aware services based on mobile agents using programmable network facilities are presented in [Winograd01], [Hightower01], [Hunt98], [Helin03a] and [Wei03]. All of them represent the context-aware research activity and implementation work by considering context location, context identity, roles or context objects as the solution to contain all context information in a model necessary to efficiently provide services, but the management of pervasive services is not well described.

3.3.1.2 Pervasive Services

Context-awareness is enabling the next generation of mobile networks and communication services to cope with the complexity, heterogeneity, dynamicity and adaptability required in pervasive applications. Context-awareness, as has been described in this chapter, refers to the capability of a system to be sensitive and react to user's and network environment, thus helping to dynamically adapt to context changes. However, context information is as complex and heterogeneous as the services that it intends to support. Furthermore, it is difficult to imagine context-aware systems that are not supported by management systems that can define, manage and distribute context efficiently. In fact, this is one of the main problems to face in the ubiquitous computing area and in pervasive applications.

It is assumed that demands for services operating over NGNs are context-aware. This means that the functionality of the service is driven by the user context, so that it can automatically adapt to changes in context. Moreover, in pervasive computing environments, the service examines both user and network context, as Dey describes in [Dey01]. Different scenarios can be envisaged that highlight the impact of context-awareness. For instance, in an emergency scenario, context-aware services will manage the incoming calls to a Voice Server, permitting only privileged calls within the emergency area. Many other examples could be offered, all of them exhibiting as a common denominator, the use of context information to provide improved multimedia services to users by adapting to different conditions and needs.

Early proposals for context-aware network services have resulted in the design and implementation of integrated platforms that cater to the full range of issues concerning the provision of LBS, such as [Henricksen02], [Komblum00]. Most initiatives propose models that refer to the situation that surrounds the user of the service as a physical person; in this user-centric model, definition of context and the context model are mainly derived from the fact that the context information is going to be used and stored only on small mobile devices used by specific users. In those proposals, the idea of using context is not flexible or extensible enough for pervasive applications, since the scope of pervasive services can cope with many different types of context-aware applications that are supported by mobile and non-mobile devices, and used by both people and virtual users (machines, other applications, etc.).

This larger scope requires the use of a more standard context format. Other initiatives aim to locate users by establishing a network of beacons located within the buildings that the users move in [CRICKET]. Another approach aimed to design services personalized to users and sensitive to their physical surroundings is [Schilit95]. Another approach is used in [LOVEUS]; this follows the client–server principle in order to adapt the delivery of sensitive multimedia information according to user's location described in [INMOVE].

The CONTEXT system [IST-CONTEXT] and all its information model representation has been designed without any preconception about the nature or type of context information that the services are going to manage. Other projects attempt similar objectives that CONTEXT has; only some of them propose the use of programmable network technologies for context-aware applications for providing services [Karmouch04], [Kanter00], [Wei03], [Klemke00], [Yang03a], [Kantar03] but not for controlling the management of services. All of these attempts develop frameworks and recommendations for communications between intermediaries with the network, especially for content peering and use for adaptation purposes. Similar initiatives could be taken to adapt the execution environment to other technologies or devices, but in this chapter, the aim is to add the functionality to manage the service operations as well.

3.3.1.3 Context Model

When humans talk with humans, they are able to use implicit situational information, or context, to increase the understanding of the conversation. This ability does not transfer well to humans interacting with computers, and especially does not work well with computers communicating with computers. In the same way, the ability to communicate context as applied to different levels of abstractions (e.g. business vs. network concepts) between applications is also difficult; in fact, the flexibility of human language becomes restricted in computer applications. Consequently, the computer applications are not currently enabled to take full advantage of the context of human–computer dialogue or computer applications from different levels of abstractions or interactions.

By improving the computer representation and understanding of context, the richness of communication in human–computer interaction is increased. This makes it possible to produce more useful computational services.

In order to use context effectively, it has to be clear what information is context and what is not. This will enable application designers to choose what context information to use in their applications, and how can it be used, which will determine what context-aware behaviours to support in their applications. Nowadays, literature in regarding to pervasive computing has outlined the benefits from using context-awareness and has proposed quite similar context-awareness definitions. Sometimes it is common to find definitions giving conceptual descriptions, which give emphasis to the influence of the environment on a process, while others with a more practical orientation try to discover the necessary of context information for a broad range of application types. Choosing the right definitions depends on the application area. Hereafter, the concept of context and context-awareness from different sources is discussed.

The first definition of context-awareness which gives origins to pervasive applications was given by Schilit et al. [Schilit95], which restricted the definition from applications that are simply informed about context to applications that adapt themselves to context. Definitions of context-awareness fall into two categories: using context and adapting to context. Further discussion on these two categories can be found in [Dey01]. In this chapter, the aim is not to re-define context or context-awareness, but rather to follow a consistent context-awareness definition that is sensitive to external variations of the information around end users, applications and networks.

Schilit and Theimer [Schilit94a] refer to context as location, identities of nearby people and objects and changes to those objects. Brown et al. [Brown96a] defines context as location, identities of the people around the user, the time of day, season, temperature, etc. Ryan et al. [Ryan97] define context as the user's location, environment, identity and time. Dey [Dey00a] enumerates context as the user's emotional state, focus of attention, location and orientation, date and time, objects and people in the user's environment. However, these definitions are difficult to apply due to their semantic diversity and not very clear applicability.

Other definitions of context have simply provided synonyms for context; for example, referring to context as the environment or situation. Some other consider context to be the user's environment, while others consider it to be the application's environment. Brown [Brown96b] defined context to be the elements of the user's environment that the user's computer knows about. Franklin and Flaschbart [Franklin98] observe it as the situation of the user. Ward et al. [Ward97] views context as the state of the application's setting. Hull et al. [Hull97] include the entire environment by defining context to be aspects of the current situations. These definitions are also very difficult to put in practice.

Pascoe defines context as the subset of physical and conceptual states of interest to a particular entity [Pascoe98]. But all the above definitions are too specific.

Finally, the definition which is given by Dey and Abowd is: "Context is any information that can be used to characterize the situation of an entity, an entity is a person, place or object that is considered relevant to the interaction between a user and an application, including the user an application themselves" [Dey00a].

From these definitions, the important aspects of context are as follows: where are you, who are you with and what resources are nearby? In an analogous fashion, it refers to a network device or the devices around it, the authors converge to define context to be the constantly changing environment, which includes the computing environment, the user environment and the physical environment, which in network terms are equivalent to the network environment, the device environment and the element environment.

If a piece of information can be used to characterize the situation of a participant in an interaction, then that information is context. For example, the user's location can be used to determine the type of services that the user receives; this interaction is context information that can be used by an application. A general assumption is that context consists only of explicit information (i.e. the environment around a person or object).

Perhaps the most important requirement of a framework that supports a design process is a mechanism that allows application builders to specify the context required by an application. In the framework of context information handling processes, it is easy to see that there are two main steps: (1) Specify what context an application needs and (2) Decide what action to take when that context is acquired. In these two steps from [Dey00a], the specification mechanism and, in particular, the specification language used must allow application builders to indicate their interest along a number of context dimensions or nature of the context addressing. However, to achieve the goal of defining an efficient specification mechanism, special interest must be taken to identify when context information is being handled with respect to other context information and other management data. Possible scenarios when context is being handled are described as follows:

Single piece of context vs. multiple pieces of context
In this example, the nature of the single piece of context is the same as the nature of the multiple pieces of context that it is related to. For example, when a single piece of context could be the location of a user and multiple pieces of context could be the location of other users.

Multiple, related context information vs. unrelated context information
Related context means different types of contextual data that apply to the same single entity. For example, related context about a user could include the location and the amount of free time. By contrast, unrelated context could be the date and time of the user and the price of a particular product.

Unfiltered vs. filtered context notifications

For example, an unfiltered request to be notified about a user's location change would result in the requesting entity being notified each time the user moved to a new location. By contrast, a filtered request might allow the requesting entity to be notified only when the user moved to a different building.

Non-interpreted context vs. interpreted context

For example, a GPS returns its location in the form of a latitude and longitude pair. If an application is interested in street-level information and no sensor provides that information, but an interpretation capability that converts latitude–longitude pairs to street name exists, then the request can be satisfied using a combination of the sensor and the interpretation. This runtime mechanism will ensure that when applications receive the context they have requested, it will not require further interpretation or analysis.

Based on previous context confrontation scenarios, context is any information that can be used to characterize one or more aspects of an entity in a specific application or network service. An entity can be a physical object such as a person, a place, a router or a physical link, or an entity can be a virtual object such as an IPsec tunnel or an SNMP agent. It is clear that context requires non-trivial managing and processing. In addition, scalable distribution and transparent storage and retrieval of this information are necessary in order for context-aware information to be effectively used.

An important requirement is that context information should not be specified in any one specific or restrictive format; rather, it should be open and available using a standardized format that can be easily supported or at least easily mapped to. This requirement drives us to use the Context Model, as it is the most general of the organization methods described.

3.4 Ontology Operations to Support Network and Cloud Service Management

Due to the proliferation of multiple services and vendor-specific devices and technologies, ontologies offer a scalable set of mechanisms to interrelate and interchange information. Therefore, basic types of ontology tools are required to support pervasive system management operations using knowledge as its inherent nature. This section reviews ontology operations and describes the ontologies features to allow service and network systems to accomplish with knowledge representation.

3.4.1 Ontology Engineering

Ontologies were created to share and reuse knowledge [Gruber93b], [Guarino95] and recently, applications have concentrated on avoiding the interoperability problems (e.g. the inability to exchange and reuse data) when different systems that use

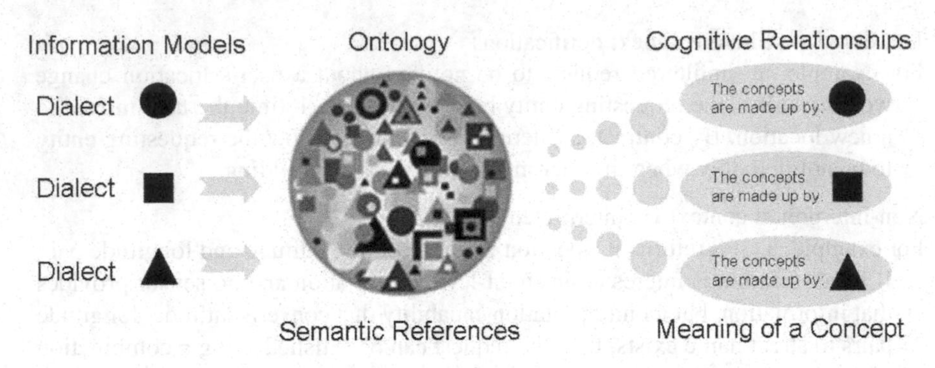

Fig. 3.2 Information models—ontology engineering

different knowledge representations and languages which interact with each other. Ontologies not only provide enrichment to the information model and semantic expressiveness to the information as described in [Gruber93b], but also allow the information to exchange between applications and between different levels of abstraction, which is an important goal of pervasive computing.

In this section, it is discussed the fact that ontologies are used to provide semantic augmentation, addressing the cited weaknesses of current management information models [López03a] and beyond with ontologies the integration of context for managing operation control is proposed, resulting in improved system management.

The cognitive relationships are shown in Fig. 3.2, where the ontologies are used for making ontological commitments in the form of cognitive relationships (i.e. an ontological commitment is an agreement to use a vocabulary in a way that is consistent to different domains of application).

The commitments are very complex [Uschold96] and can be thought of as a set of mappings between the terms in an ontology and their meanings. Hence, ontologies can be combined and/or related to each other by defining a set of mappings that define precisely and unambiguously how one concept in one ontology is related to another concept in other ontology.

In most current management applications, different data models are embedded in each application, and as a result complex systems need to be developed to translate between data defined by different applications. This is due to many reasons; perhaps the most important is different management applications require different management data to accomplish different tasks or to represent information from a different point of view.

Very often, each application uses different tools, since the use and manipulation of those data require different functions. For example, the simple text-based functionality of LDAP (for directories) is not sufficient for more complex tasks that require (as an example) SQL. This is the trap which developers fall into when they use an application-specific data model instead of an application-independent information model. Furthermore, the complexity increases when end-user applications use information models that need to interact with information models from devices in the networks, as the difference between user and network data is significant.

3.4.1.1 Ontology Engineering Definition

Ontology is a "formal mechanism for representation of a conceptualization in a shared domain". This definition was defined in [Gruber95]. Gruber defines an ontology as a description (like a specification of a program in a formal language) of the concepts and relationships that can exist for an entity or a set of entities using a formal representation.

An ontology must be explicit, formal and open. Explicit means that the entities and relationships used, and the constraints on their use are precisely and unambiguously defined in a declarative language suitable for knowledge representation. Formal means that the ontology should be represented in a formal grammar. Open means that all users of an ontology will represent a concept using the same or equivalent set of entities and relationships.

In terms of practicality and usage, an ontology is not only for knowledge representation. For example, multiple researchers [DeBruijn03], [Keeney06], [López03b], [Horrocks04] show many advantages of using ontologies in the IT area, such as for capturing, defining, sharing and reusing knowledge, along with verifying the correctness of knowledge and being able to reason about an event using the stored knowledge of the ontology.

In the context of knowledge engineering, and based on the definitions and concepts from [Gruber95] and [Guarino95], it is easy to observe that this cited definition is consistent with the usage of ontology as a set-of-concept-definitions, in a general form. However, when concepts about management and context information are being integrated, the knowledge representation is more than a set of definitions; it requires more semantic enrichment (references between the different and diverse domains) for building up the cognitive relationships. So a newer definition from [Serrano07a], [Strassner07], considering those requirements, is supported and used with more practical value in communications:

> An ontology is a formal, explicit specification of a shared, machine-readable vocabulary and meanings, in the form of various entities and relationships between them, to describe knowledge about the contents of one or more related subject domains throughout the life cycle of its existence… Formal refers to the fact that the ontology should be represented in a formal grammar. Explicit means that the entities and relationships used, and the constraints on their use, are precisely and unambiguously defined in a declarative language suitable for knowledge representation. Shared means that all users of an ontology will represent a concept using the same or equivalent set of entities and relationships. Subject domain refers to the content of the universe of discourse being represented by the ontology.

The use of ontologies for the integration of context information is becoming a common practice. In the framework of this book ontologies are used to represent management operations. Concepts from autonomic communications systems about information interoperability between systems are used, and more particularly regarding the free exchange of information between resources, systems and other applications and other systems.

3.4.1.2 State of the Art in Ontology Engineering

Not all ontologies are built using the same language and structure. For example, Ontolingua uses the knowledge interchange format (KIF) language and provides an integrated environment to create and manage ontologies (more details about KIF can be found in [Genesereth91]). KL-ONE [Brackman85], CLASSIC [Borgida89] and LOOM [Swartout96] each use their own ontology language. The open knowledge base connectivity (OKBC) language, KIF and CL (common logic) have all been used to represent knowledge interchange, and have all become the bases of other ontology languages. There are also languages based on a restricted form of first-order logic (this makes the logic more easily computable), known as description logics, such as DAML + OIL [Horrocks02].

With the advent of Web services, a new family of languages appeared. The resource description framework (RDF) [Brickley03a] and RDF-Schema [Brickley03b] have provided basic ontological modelling primitives, like classes, properties, ranges and domains. RDF influenced the defense agent markup language (DAML) from the USA [DAML]; DAML and OIL (the ontology inference layer, a separate but parallel European effort) [Horrocks02] were eventually merged in the World Wide Web Consortium (W3C), which created the Web ontology language (OWL) standard [OWL] and it was introduced in [Dean02]. OWL is an integral part of the semantic Web [Berners-Lee01] and a W3C recommendation [W3C]. OWL comes with three variations: OWL Full, OWL DL and OWL Lite [DeBruijn03]. OWL Lite has been recently extended to create OWL-Flight, which is focused on using a logic programming framework [DeBruijn04]. Other activities are inspired by first, integrating semantic rules into an ontology (this effort is inspired by some OWL modelling weaknesses to contain certain restrictions) and second, building new languages on top of OWL for specific applications. The best example of this is OWL-S, which was designed to be used with semantic Web applications [OWL-S]. Another approach is SWRL (semantic Web rule language) combining sublanguages of the OWL (OWL DL and Lite) with those of the rule markup language (unary/binary datalog) [Horrocks04].

Referring to information integration, some initiatives are based on a single global ontology, such as TSIMMIS, described in [Garcia97]. Another example is the information manifold in [Kirk95]. Others use multiple domain ontologies, such as InfoSleuth [Bayardo97] and Picsel [Reynaud03], but any of them could be adapted for integrating and gathering context information for various service applications in autonomic environments.

The work above is focused on using ontologies for knowledge engineering representation. Ontologies have also been used for representing context information in pervasive applications. In particular, the CoOL (context ontology language) is an initiative for enabling context-awareness and contextual interoperability as described in [Strang03c]. CoOL allows context to be expressed, which enables context-awareness and contextual interoperability, but does not describe how to manage context or context-aware services.

The friend of a friend (FOAF) ontology [Brickley03a] allows the expression of personal information and relationships, and it is a useful building block for creating information systems that support online communities. This ontology does a good job of describing personal profiles, but has very little user and virtually no network context information.

The standard ontology for ubiquitous and pervasive applications (SOUPA) [Chen04] is a standard ontology for ubiquitous and pervasive applications, and includes modular components with vocabularies to represent intelligent agents with associated beliefs, desires and intentions. SOUPA describes only the information necessary for intelligent agents, and does not define any specific management services.

The context broker architecture ontology (COBra-ONT) and MoGATU BDI ontologies aim to support knowledge representation and ontology reasoning in pervasive computing environments, each one in specific tasks. While COBra-ONT [Chen03a] focuses on modelling context in smart meeting rooms, MoGATU BDI [Perich04] focuses on modelling the belief, desire and intention of human users and software agents. Without any doubt, one of the most interesting works in the integration of context towards pervasive services is COBra-ONT [Chen03a]; however, in that proposal, the importance of context-awareness for management operations is marginal, whereas it is important to integrate the context information as part of the network life cycle operations.

The PLANET ontology [Gil00] was created for representing plans with a more service-oriented vision. PLANET aims to be a comprehensive ontology that can be reused to build new applications by using the knowledge from business and then translate that knowledge to computing systems, describing several specializations of it to represent plans in different real-world domains. Planet is available in Loom, KIF and CycL, but not available in standard languages like OWL.

The CONON ontology [Wang04] is an OWL-encoded context ontology for modelling context in pervasive computing environments, and for supporting logic-based context reasoning. CONON provides an upper context ontology that captures general concepts about basic context and also provides extensibility for adding domain-specific ontology in a hierarchical manner. CONON solves the problem of extensibility by using standard DAML + OIL or OWL. However, as it is context-centred, its management operations are restricted.

Recently, other approaches more context- and business-oriented in the framework of research projects have been created such as the COMANTO ontology [Roussaki06] in the framework of the European DAIDALOS project [DAIDALOS]. COMANTO is an ontology for describing context types and inter-relationships that are not domain-, application- or condition-specific. The objective is just to augment services with context-awareness functionality. This non-application dependence is a limitation when knowledge from two different domains is required to be integrated.

Many developed and/or currently being developed benefit from the concept of autonomic computing, and follow the premise of optimizing the support for user-oriented services using operational and/or business support systems (OSS and BSS).

In these complementary scenarios, the use of the ontologies is more than just the "simple" representation of knowledge; rather, ontologies are also used to integrate knowledge. Finally, the CONTEXT project [IST-CONTEXT], which acts as the base for extending the information model and formalize it with ontologies, defines an XML Policy Model that supports the complete service life cycle. The policy model is extensible and contains parts defined as context information. This approach follows the business-oriented scope based on context information that the networks require to operate. The service life cycle is managed by a set of policies that contain such context information, and it is used to trigger events. However, this proposal does not use appropriate formalisms for sharing context information for supporting the reuse of context information contained in the policies.

Ontology is used for expressing different types of meaning for a concept that needs to be interpreted by computers. There are ontologies aiming not only to define vocabulary to enable interoperability, but also to define one or more definitions and relationships for a concept. This feature enables different applications to use different meanings for the same object in multiple applications, which helps integrate the cross-layers in NGN systems. Due to the inherent influence of the Internet, most initiatives for representing context information want to use schema extensions that support Web services and other initiatives specified by the [W3C].

3.4.1.3 Ontology Engineering Applied to Network and Service Management

An important characteristic of ontologies is their capability to share and reuse information. This reusability is the feature that attracts the attention of many developers of information systems and obviously this feature is applicable to communications systems, particularly in management domains. Sharing and reusing information depends on the level of formalism of the language used to represent information in the ontology.

One way to share and resource network knowledge is to use models and structures which are extensible enough to enable such information to be captured [IBM01b], [Kephart03]. Initiatives for using ontologies in the domain of networking are [Keeney06], [López03a], [Guerrero07]. More specifically, in the pervasive services area, context information is essential and could be used for managing services and operations [Strassner06a]. However, it is a highly distributed environment and introduces a great challenge to managing, sharing and exchanging of information in a consistent and coherent manner. To do so, where current networking scenarios use different networks, technologies and business rules and a diverse interaction of domains increase the complexity of the associated management activities, the emerging of autonomic solutions is acquiring importance.

In autonomic environments, mechanisms are necessary for managing problems in an automated way, minimizing human interaction, with the objective of handling problems locally. In autonomic environments, every technology uses its own protocols, and most of the time proprietary languages and management data structures, so the interoperability for exchanging information is impaired. Autonomic environments

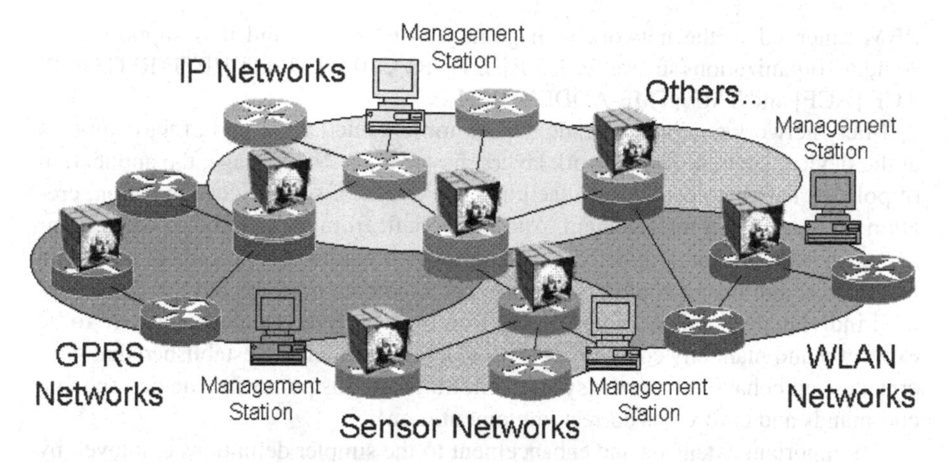

Fig. 3.3 Multiplicity of technologies in autonomic environments

seek to unite these isolated stovepipes of data and knowledge using semantic mechanisms to share and reuse information. This often takes the form of middleware that understands and translates information, commands and protocols.

Figure 3.3 depicts autonomic environments, a diversity of technologies involved in the exchanging of information, which results in increased complexity [Serrano07b]. It is observed that every management systems or station corresponds to every technology domain; in autonomic communications, the exchange of information implies the collection and processing of the information by using the same information model. So the systems must have the necessary mechanisms to translate the information into the same format that the model define.

3.4.2 Autonomic Computing and PBM

A policy has been defined as a rule or a set of rules that manage and provide guidelines for how the different network and service elements should behave when certain conditions are met [IETF-RFC3198]. Verma defines a policy as a directive that is specified to manage certain aspects of desirable or needed behaviour resulting from the interactions of user, applications and existing resources [Verma00]. However, as said earlier in this chapter, as reference the definition of "Policy is a set of rules that are used to manage and control the changing and/or maintaining of the state of one or more managed objects" is used [Strassner04].

The main benefits from using policies are improved scalability and flexibility for managing services. Flexibility is achieved by separating the policy from the implementation of the managed service, while scalability is improved by uniformly applying the same policy to different sets of devices and services. Policies can be changed dynamically, thus changing the behaviour and strategy of a service.

PBM emerged in the network management community, and it is supported by standard organizations such as the IETF [IETF-RFC3060], DMTF [IETF-RFC3460], ACF [ACF] and TMF [TMF-ADDENDUM].

Many network equipment vendors have implemented all or part of these models at the device, element and network layers; however, in NGN usage, the application of policies is being abstracted to facilitate the works of service customization, creation, definition and management. Another benefit from using policies when managing services is their simplicity. This simplicity is achieved by means of two basic techniques: automated configuration (e.g. each element does not have to be configured individually) and simplified abstraction (e.g. each device does not have to be explicitly and manually configured—rather, a set of policies is established that governs desired behaviour, and the system will translate this policy into device-specific commands and enforce its correct implementation).

An important extension and enhancement to the simpler definitions employed by the IETF and DMTF that is very attractive to autonomic systems has been proposed [Strassner04]. This extension is currently being implemented by the ACF. Specifically, policies are linked to management using finite state machines. The policy definitions studied in this section has been defined in [Strassner04], [Strassner06a], which are along discusses in this book:

> Policy is a set of rules that are used to manage and control the changing and/or maintaining of the state of one or more managed objects.
>
> A PolicyRule is an intelligent container. It contains data that define how the PolicyRule is used in a managed environment as well as a specification of behaviour that dictates how the managed entities that it applies to will interact. The contained data is of four types: (1) data and metadata that define the semantics and behaviour of the policy rule and the behaviour that it imposes on the rest of the system, (2) a set of events that can be used to trigger the evaluation of the condition clause of a policy rule, (3) an aggregated set of policy conditions, and (4) an aggregated set of policy actions.

Policy management is expressed using a language. Since there are many constituencies having their own concepts, terminology and skill sets that are involved in managing a system (e.g. business people, architects, programmers and technicians), one language will not be expressive enough to accommodate the needs of each constituency. Figure 3.4 shows the approach used in [Strassner06a] and termed as policy continuum, defined in [Strassner04] and extended in [Davy07a]. While most of these constituencies would like to use some form of restricted natural language, this desire becomes much more important for the business and end users.

In the framework of this book, the definition of a language, following principles from policy continuum, supports the idea of an initial representation by using XML to ensure platform independence. In the same way, implemented dialects are also easy to understand and manage, and the large variety of off-the-shelf tools and freely available software provide powerful and cost-effective editing and processing capabilities.

Each of the implementation dialects shown in Fig. 3.4 is derived by successively removing vocabulary and grammar from the full policy language to make the dialect suitable for the appropriate level in the policy continuum. XML representations and vocabulary substitution by using ontologies is applied or some view levels in the policy continuum.

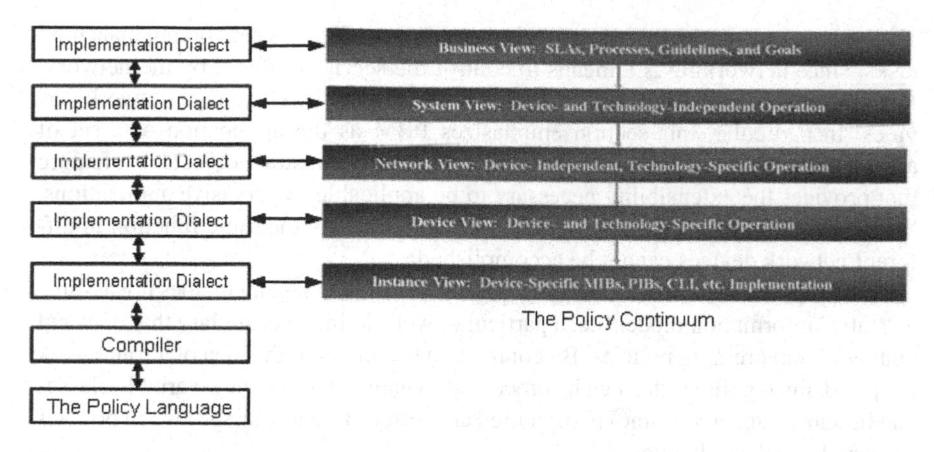

Fig. 3.4 Mapping of policy language dialects to the policy continuum

3.4.2.1 Policy Management Applied to Network and Service Management

The main objective of using policies for service management is the same as that of managing networks with policies to automate management and do it using as high a level of abstraction as possible. The philosophy for managing a resource, a network or a service with a policy-based managed approach is that "IF" something specific happens "THEN" the management system is going to take an action. The main idea is to use generic policies that can be customized to the needs of different applications; the parameters of the conditions and actions in the policies are different for each user, reflecting its personal characteristics and its desired context information. The asset idea is the use of the policy-based paradigm to express the service life cycle and subsequently manage its configuration in a dynamic manner. It is this characteristic which provides the necessary support and operations of pervasive systems.

The policies are used in the management of various aspects of the life cycle of services. An important aspect of policy-based service management (PBSM) is the deployment of services throughout the programmable elements. For instance, when a service is going to be deployed over any type of network, decisions that have to be taken in order to determine which network elements the service is going to be installed and/or supported by. This is most effectively done through the use of policies that map the user and his or her desired context to the capabilities of the set of networks that are going to support the service. Moreover, service invocation and execution can also be controlled by policies, which enable a flexible approach for customizing one or more service templates to multiple users. Furthermore, the maintenance of the code realizing the service, as well as the assurance of the service, can all be related using policies. A final reason for using policy management is that when some variations in the service are sensed by the system, one or more policies can define what actions need to be taken to solve the problem.

The promises of PBM are varied and often conceptualized as networking managing tasks, since networking is a means to control the services offered by the network. However, policies can potentially do much more than "just" manage network services. In particular, this section emphasizes PBM as the application of a set of abstract condition–action rules. This ability to manage abstract objects is a feature that provides the extensibility necessary to be applicable to pervasive applications. Without this ability, a common interface to programming the same function in different network devices cannot be accomplished.

The proposed use of the PBM paradigm for service management does not assume a "static" information model (i.e. a particular, well-defined vocabulary that does not change) for expressing policies. By contrast, in this chapter, the idea of a framework for pre-defined policies that can be processed dynamically (e.g. new variable classes can be substituted at runtime) is supported as it offer more advantages when policies are pre-defined and known.

Associations between the information expressing the policy structure, conditions and actions with information coming from the external environment are crucial to achieve the goals of management systems. Specifically, the externally provided information can either match pre-defined schema elements or, more importantly, can extend these schema elements. The extension requires machine-based reasoning to determine the semantics and relationships between the new data and the previously modelled data. This is new work that augments previous PBM systems, and is assumed to reside outside the proposed framework (the service creation and customization systems in the context system).

By supporting dynamically pre-defined policies, the flexibility of pervasive management can be achieved and context interactions can be more completely realized using policy-based control. This feature is a requirement of the design of the overall pervasive system (for achieving rapid context-aware service introduction and automated provisioning).

3.4.2.2 State of the Art of PBM

PBM has been proven to be a useful paradigm in the area of network management. In the last few years, initiatives have appeared that use polices or rule-based decision approaches to tackle the problem of fast, customizable and efficient service delivery. Among the most representative are OPES by Piccinelli [Piccinelli01] and Tomlinson [Tomlinson00].

This analysis goes a step further to analyze solutions intended to control the full service life cycle by means of policies. To accomplish this goal, there are solutions making use of programmable network technology, for example as the technology infrastructure supporting pervasive services and applications. Programmable technology as described in [Raz99] plays the role of the infrastructure supporting context-aware applications and services while, at the same time, it support networking operations to guarantee the correct operation of the network. The IST-CONTEXT project approach [IST-CONTEXT] and the ANDROID project [ANDROID] aims to prove the feasibility of providing a managed, scalable, programmable network infrastructure.

The outcomes when adaptive network technology and PBM interact to exchange and reuse information are another innovation. As it represent sharing process of information from middleware approach to communication networks. Programmable network technology endow to the network with the capability to process certain applications for executing specific management operations with a level of self-decision [Raz99].

This programmable characteristic can also play a key role in the management of context information; if the information can be send in a format according to the information model. In other words, besides main programmable technology functionality concerning the support of the service code, the programmable network can be seen as a technological support for the actual service management layer in the sense that the latter can extend its functionality as needed by sending the appropriate network instructions, based on the context information acting as a trigger to the programmable network nodes running anywhere, and execute it.

There are important approaches and projects dealing with PBSM [Jeng03] and [Joshi03], and each one offers a policy specification for controlling the service management operations in some cases. A survey of policy specification approaches has been provided in [Damianou02]. The most important goal is to have a broad panorama of solutions achieving the interaction between information systems handling the context information from users and networks and the technological infrastructures supporting network services.

To date, most of the approaches deal with introducing the context information in networking tasks [Sloman94a] and [Sloman99c], leaving or ignoring services, where the most added values or business opportunities can be created. The more symbolic approach is PONDER as an object-oriented and declarative language, mainly adopted for object-oriented distributed systems [Damianou01].

Traditionally, PBM systems does not impose constraints to define management policies that are dynamically operated, in fact those policies can be defined without considering the behavior of a system at all. This non-dependence between system behavior and policy operation is an advantage from one side, because it does not set restrictions to the service control while systems is being operated (e.g. changing code operation does not require the cooperation of components being governed). However, in the other hand this non-dependency generates that new services cannot be configured dynamically because the necessary information to activate the new service is not implicit in the policy information model. As a consequence the new service cannot be related with user profiles, devices and systems information cannot be related with user profiles, device and systems information.

Another proposal is the CONTEXT project [IST-CONTEXT], the objective of which is to provide a solution in the form of an architecture that is used for creating, deploying and managing context-aware services and relates user and behaviour of a system to information for triggering the configuration of new services. This approach is well oriented for service management applications under the scope of policy-based systems, but was developed without considering the multiplicity of context information and their technology dependence.

Generalized PBSM architectures for autonomic systems will articulate a functional process model and include process inter-relationships for an organization. It shall encompass a methodology that identifies the necessary policies, practices, procedures, guidelines, standards, conventions and rules needed to support the business and their process inter-relationship. This enables the organization to govern the application of policy management mechanisms to appropriate managed entities. Hence, the application of the policy-based paradigm to the service management area seems to be in a good way to meet next-generation service management goals encompassing user and business goal offering a tool for dynamic service provisioning.

Some other initiatives include ontologies for their formalism process for creating a policy language which is able to be used in PBSM; in those initiatives the formalism with formal languages brings the advantage of defining information models with the objective of sharing the information as a form of knowledge at different application and networking levels. This orientation comes from the necessity and requirements to satisfy the increasing tendency towards creating solutions based on semantic Web for supporting Web services and applications. KAoS is proposed as a policy language for domain Web services and it propose policy specification, disclosure and enforcement of policies for Web services, which the semantic has control over some features of the service [Uszok04].

Another work in the context of Web services being supported by policies is Rei [Kagal03]. Rei concentrates on how to describe policy operations of their functional stages defining a language which is able to support the semantic descriptions present in Web services. Rei is composed of several ontologies, ReiPolicy, ReiMetaPolicy, ReiEntity, ReiDeontic, ReiConstraint, ReiAnalysis and ReiAction [Kagal02]. Every ontology describes classes and properties associated with their unique namespace definition or domain. A comparison study offering a general scope of these approaches is presented in [Tonti03], even in this study ponder language is included as referent policy language in the field of networking and study its possibilities to support Web services.

3.4.2.3 Policy Information Model to Support Network and Service Management

The promises of PBM are varied and today demonstrated as suitable for supporting network operations and services. Most approaches to management and network configuration lack the capacity for exchanging and reusing information from business goals and technical objectives, as they cannot relate network services to business operations. Furthermore, these conditions avoid new business roles for modifying the systems in terms of adapting services to the demands of changing user needs and environmental conditions.

A typical example is understood when traditional management protocols are unable to express business rules, policies and processes in a standard form (e.g. SNMP or CLI). They have no concept of a customer, and hence when they report a fault, it is impossible to determine which, if any, customers are affected from data retrieved by the protocol or even the commands. This makes it nearly impossible to use traditional management protocols to directly change the configuration of the network in response to new or altered business requirements. Instead, software solutions must be used that translate the business requirements to a form that can then be realized as SNMP or CLI instructions.

In addition, the software solutions must be designed and implemented using wrappers that allow them to be used in conjunction with management data from the diversity of fixed and wireless technologies that have countless variations of information, formats and meanings. Thus, the problem is how to represent the appropriate semantics associated with the context information, so that the context information can be used to govern the management operations. To do so, the use of semantic tools as formal mechanisms is applied.

A policy information model that has the capability to use context information provides a cohesive, comprehensive and extensible means to represent things of interest in a managed environment [Opperman00]. Things of interest are the inclusion of users, policies, processes, routers, services and even protocol configuration (cross-layered interoperability). Extensibility is achieved using a combination of software patterns and abstraction mechanisms, such as events which, for example "transport" the information through the components of a system.

Information models are the glue that enables different components, manufactured by different vendors, to interoperate—whether they are network devices, software or a mixture of them. As an abstraction and representation of the entities in a managed environment, the attributes, operations and relationships of managed entities are studied in this section, independent of any specific type of repository, software usage or access protocol defined in the information model.

As an approach to this, the directory enabled networks for NGNs (DEN-ng) has been studied. An assumption is that the use of an information model facilitates the description of the business, system, implementation and runtime aspects of managed entities and their relationships. This is because it can be used to represent each of these aspects, and thus associate one aspect with the others. Information models excel at representing only the detail required for only the entities of concern in a management domain. Information models use a well-defined taxonomy, meaning that all managed entities can be inherently related to each other using the information model.

A simplified version of the preliminary DEN-ng context model is shown in Fig. 3.5. This context model is unique, in that it relates context to policy to management information [Serrano07a], [Strassner07]. Conceptually, this model works as follows: context determines the working set of policies that can be invoked; this working set of policies defines the set of roles and profiles that can be assumed by the set of ManagedEntities involved in defining context. The working set of policies also defines the set of management information that is of interest (for that specific context).

The *SelectsPolicies* aggregation defines a given set of *Policies* that must be present to support the behaviour of that particular context. It is an aggregation to show that context and policy are strongly related (a whole-part relationship). This enables changes in context to change the set of policies that are used for orchestrating system behaviour.

The association *PolicyResultAffectsContext* enables policy results to influence context. For example, the success or a failure of the execution of a policy can affect the state

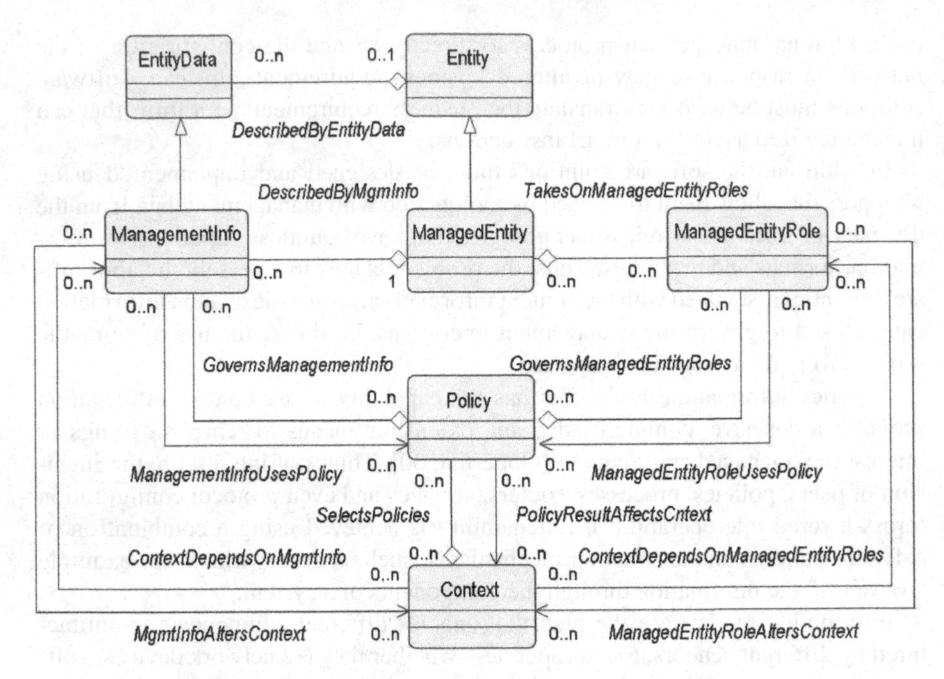

Fig. 3.5 Simplified DEN-ng context model

of the system. The selected working set of policies defines the appropriate roles of the *ManagedEntities* that form context; this enables context to manage system functionality (through roles) at a higher level of abstraction. In particular, this means that *policy determines the set of roles that can be assumed for a given context*. This is represented by the *GovernsManagedEntityRoles* aggregation. When these *ManagedEntityRoles* are defined, they are then linked to context using the *ContextDependsOnManagedEntityRoles* association; the *ManagedEntityRoleAltersContext* and the ManagedEntityRoleUsesPolicy associations are used to feedback information from ManagedEntityRoles to context and policy, respectively.

Context also defines and depends on the management data collected from a *ManagedEntity*. First, policy is used to define which management information will be collected and examined (via the *GovernsManagementInfo* aggregation); this management information affects policy using the *ManagementInfoUsesPolicy* association. Once the management information is defined, then the two associations *ContextDependsOnMgmtInfo* and *MgmtInfoAltersContext* codify these dependencies (e.g. context defines the management information to monitor, and the values of these management data affect context).

Given the above definitions, the relationship between policy and context becomes clearer. When a context is established, it can select a set of policies that are used to govern the system. The governance is done by selecting an appropriate set of *ManagedEntityRoles* that provide access to the functionality of the *ManagedEntity*. These *ManagedEntityRoles* provide control points for functionality that needs to be governed.

Similarly, the result of executing a policy may alter context (e.g. an action did not succeed, and new corrective action must be taken; or a set of configuration changes did succeed, and the system is back in its desired state) such that a new context is established, which in turn may load a different set of policies.

3.5 Ontology Engineering Functions and Tools

Ontology engineering as a mechanism for helping the computing systems to integrate knowledge has a large number of example applications, since the most simple to represent information until the most complex and robust as a complete information management system in a communication network, where diverse mechanisms and reasoning process are present.

Ontology engineering is being more accepted and considered every day as a suitable alternative to cope with one of the main problems in the communications area (to endow the communications networks with the necessary semantic enrichment to support applications and services).

The following sections in this chapter show some of the most important applications that ontologies offer to the computing systems as those are divided into two basic types of ontology tools: The first group contains the ontology mapping and merging tools, for either combining multiple ontologies. An application example in this group is when it is necessary to identify nodes into a network defined by a specific ontology (ontology A), which are semantically similar to nodes in other network defined by other different ontology (ontology B). In this example mapping operations and merging for combining concepts are necessaries. In the second group, the development of mechanisms, for creating, editing and specifying tools that can be queried using one or more inference engines are necessary.

Ontologies are used to define a lexicon that all other system components must follow and furthermore the relationships that exist between each others. Once the ontology has been created, the ontology can be handle using ontology merging tools, such as PROMPT [PROMPT] and Chimaera [CHIMAERA]. These approaches provide a common set of definitions and relationships for data used in the system based on ontologies, and are a basis for the semantic rules that are used in cognitive systems. By other side, common ontology development tools, such as Protégé [PROTÉGÉ] and Ontolingua [ONTOLINGUA], are used to define queries, commands and assertions used this kind of systems.

3.5.1 Ontology as an Operational Mechanism

Ontologies were created to share and reuse knowledge, and a formal application of these concepts can be studied in [Genesereth91], where the information is transformed using a specific KIF, even if the input data uses multiple heterogeneous representations. This is why many knowledge engineering efforts are using ontologies to specify and share knowledge. Network management problems need this capability, especially since there is a lack of a standard mapping between different languages used to represent network management data (e.g. CLI and SNMP).

The challenge is to create the links between different structural representations of the same information. The lack of using standard information models and the resulting mismatch of data models used to represent network management and context data are the motivations to use the set of ontology operational mechanisms that enable such information exchange and the interactions between different application- and domain-specific data models.

Following the premise that ontologies can be used as a mechanism for reusing and exchanging knowledge in pervasive systems for context-aware information, as explained in [Giunchiglia93], ontologies are used as operational mechanisms to provide such features to the service management systems by using autonomic-like behaviour and functionality.

Autonomic systems benefit the features that ontologies provide to the information, when the knowledge in these application- and domain-specific data models is standardized by using ontologies to provide a formal representation and a set of mapping mechanisms, autonomic systems benefit from the features that ontologies provide to the information, when it is being formalized and then transformed in knowledge, thus autonomic systems must be able to understand and use the semantics of these data, as explained in [Kitamura01].

Once the information has been gathered, the next step is for each component to make decisions based on and/or following a set of ontology-based reasoning procedures [Keeney05], which allow it to create and execute suitable inferences and/or deductions from the knowledge expressed in the knowledge base, and/or simply transfer the necessary information to different abstraction layers in autonomic systems, for example with certain level of pragmatism as result of a decision. The ontology-based procedures can be categorized in three processes described as follows.

3.5.1.1 Ontology Mapping or Alignment

Ontologies are used to describe and establish semantic commitments about a specific domain for a set of agents, with the objective that they can communicate without complicated translation operations into a global group. Examples of those commitments are presented in [Crowcrof03]. The idea of semantic commitment can be thought of as a function that links terms of the ontology vocabulary with a conceptualization. Those "agreements" can represent links between concepts from different domains or concepts from the same domain, as is exemplified in [Khedr03].

In particular, ontologies enable the system to describe concepts involved in the applications, process or tasks (a domain of discourse) without necessarily operating on a globally shared theory. Knowledge is attributed to agents that do not need to know where the commitments were done; all they need to know is what those commitments are, and how to use them.

An agent "knows" something if it acts as if it had and understood that information, so that it can act rationally to achieve its goals. Then, conditions that agents can use to operate with "actions" of the agents can be defined; this can be seen as a functional interface to tell the agents how to operate for sharing, reuse, verification and reasoning. An application of these concepts focused on communications using context concepts can be studied in [Khedr02].

A slightly different viewpoint is "an ontology mapping process is based on the identification of similar concepts present in the ontologies to be aligned, and then if those similarities exist a merging process is valid", as described in [López03c]. The alignment of ontologies, then, consists of the definition of agreements between two or more ontologies, where an agreement is a link that exists between two (or more) concepts in the ontologies. These agreements then allow the exchange of information between applications at the same and/or different levels of abstractions that have created the agreements.

The semantic commitments defined in the ontologies are used to delineate in each case the knowledge that can be shared with agents that commit to the ontologies. Likewise, the ontologies provide the semantic structures necessary to allow gathering, managing and storing efficiently context information in services and applications.

3.5.1.2 Ontology Merging or Fusion

The research activity when merging ontologies is broad. An algebra for ontology commitments has been defined [Mitra00], which uses a graph-oriented model with values for defining areas of interest and logic operations between the concepts in the ontologies such as unions, intersections and differences. The fusion of ontologies will result in the creation of a new ontology based on the set of ontologies that are being fused, as has been presented in [López03b]. In this kind of ontology merging process, all of the concepts and relationships are replaced by a new set of concepts and relationships that are equivalent to the original ontologies [López03a].

3.5.1.3 Ontology-Based Reasoning

A mechanism that is well accepted by the network management community for automated control and service composition is artificial intelligence (AI). However, formal AI-based reasoning methods are still under analysis. A general approach for describing how AI can be applied in this domain is as follows: ontology-based reasoners can be used to operate on generic as well-specific concepts by reasoning about the various states of an entity and possible changes to them.

The idea in ontology-based reasoners is to enable adaptive systems to process the decisions for ontology-based semantic elements automatically. In other words, take the functional properties of a service as inputs, and output events and actions that are dependent on the context state of the user, application and/or service using the context information and other knowledge that is available in semantic descriptions for the operation of the services, the goals of the business, user and application profiles and other environmental variables.

This vision of the context representation and provisioning using AI mechanisms enables the solution of a large range of problems. Some important notes and comments in this respect can be found in [O'Sullivan03] and [Lewis06], where the ontologies are being considered to be used for composing services in an intelligent manner, and also for providing semantic interoperability in pervasive environments.

3.5.2 Ontology as a Specification Mechanism

In general, an information or data model may have multiple ontologies associated with it. This is because ontologies are used to represent relationships and semantics that cannot be represented using informal languages, such as UML. For example, even the latest version of UML does not have the ability to represent the relationship "is-similar-to" because it does not define logic mechanisms to enable this comparison. Note that this relationship is critical for heterogeneous end-to-end management services and inclusive systems, since different devices have different languages and programming models, which means that when the same high-level directive is used, its meaning and side effects can be different [Strassner06a].

3.5.2.1 Ontology Editors

Ontologies define the lexicon that a language uses to define the set of queries, commands and assertions that are available when the ontology language is being used. Pragmatically, the language represents an agreement to use the shared vocabulary in a coherent and consistent manner. Hence, the first and most basic activity that is done with ontologies is the definition of knowledge that can be retrieved. This includes things, objects, activities and other entities of interest, including events that have occurred in the environment of the system, as well as relationships between these entities. This enables different sensor elements, such as agents, to all use the same formal language to describe contextual data in a common way.

Today, the most common exemplar for a service definition language is without any doubt the semantic Web. The huge quantity of information on the Web emphasizes the need to have a common lexicon, which in turn raises interest in using ontologies.

The semantic Web gave rise to a new family of languages, including the RDF and the OWL standards. Both are integral parts of the semantic Web; the latter depends on the former and is also a W3C recommendation. OWL comes with three variations (OWL Full, OWL DL and OWL Lite), as described in detail in the state-of-the-art section in Chap. 2. Each of these different dialects has its own strengths and weaknesses, and provides different levels of expressiveness for sharing knowledge.

In the other hand, when designing an ontology, it is very common see ontology languages providing their own GUI to edit the ontology. Many of those tool very powerful, however it usually reflect the intent of the designer of the ontology language, and hence may or may not be applicable for a particular application domain. Hence, open source alternatives exist that work with standard languages (such as the OKBC [OKBC] or KIF standards [KIF]). In order to maximize reuse, the research activity has been done using open source tools that are not dependent on any one specific commercial product to edit the ontologies.

3.5.2.2 Ontology Reasoners

Technical aspects of the OWL language, and particularly OWL DL, have its foundations in descriptions logics, which is fundamentally a subset of first-order logic. An inherent property of first-order logic is that it can be described and contained in algorithms described in a finite number of sequences or steps, but do not guarantee the result that the steps will be completed in finite time.

A first internal evaluation of using ontologies that use reasoners is based on the flexibility of the various inference services offered that can be used to determine the consistency of the ontology. A class is inconsistent when it cannot possibly have any instances. The main inferences services can be listed as follows:

1. Inferred superclasses of a class.
2. Determining whether or not a class is consistent.
3. Deciding whether or not one class is subsumed by another.

An ontology reasoner is normally used for verification of the ontology; however, other more extensive uses focused on generating solutions and creating decisions (decision-making processes). Hence reasoner plays an important role when using ontologies and it is necessary to identify the correct one to increase the efficiency according to the compatibility with an ontology editor.

3.6 Conclusions

In this chapter…

Reference to former research challenges in ontology-based principles for network management, in order to build a clear framework describing how ontologies can be used to represent context information in network management operations, has been introduced and discussed.

The main benefits from using policies for managing services are improved scalability and flexibility for the management of systems; this feature also simplifies the management tasks that need to be performed. These scalability and simplification improvements are obtained by providing higher level abstractions to the administrators, and using policies to coordinate and automate tasks.

Policies make automation easier, as well as enable service management tasks to take into account any customization of the service made by either the consumer or the service provider. PBM models have been studied due to its use of patterns, roles and additional functionality not present in other models (e.g. integration and support of context information as events in the actions and events of a state that is managed by policy).

The extension of service management functionality to act on demand is an important property supported by using PBSM systems. The current state of the art in this area has been studied and discussed from two important points of view. First, a novel combination of policy management and context to describe changes to service management functionality and, more importantly, how to respond to these changes has been analyzed. Second, the main management operations for controlling the service life cycle in most of the pervasive services have been defined. As a general conclusion, this chapter is a guideline about policy based management systems, and ontology engineering to build a systematic language by using formal data representations and by using modelling information techniques.

The modelling techniques applied on policy and information management can be used to build a componentized inherently scalable information infrastructure architecture for managing pervasive services as will be described in a subsequent chapter in this book.

Chapter 4
Information Modelling and Integration Using Ontology Engineering

4.1 Introduction

Ontology engineering has been proposed as a formal mechanism for both reducing the complexity of managing the information needed in network management and autonomic systems and for increasing the portability of the services across homogeneous and heterogeneous networks. This section describes a formal mechanism to integrate context information into management operations for pervasive services.

Reuse of existing ontologies is a task that ontology development must anticipate. In fact, it is this feature that will speed up the development of new extensible and powerful ontologies in the future. The integration of ontologies is an important task in the ontology development area. Hence, it is important to emphasize that the perspective and analysis of the ontology is crucial in its level of acceptance. In this sense, in this chapter there is a devoted section to present a description and representation of a practical example to represent an ontology and integrate concepts from different domains, the ontology is no implemented as full but the formal representation and the conceptual background to build the ontology are fully explained.

In this chapter, an ontology model definition and representation is introduced and developed, following the formal methodology basic principles described in this book the ontology is explained in a conceptual form. It is a demonstrative application to integrate user's context information in service management operations. This ontology provides the semantics, using a certain level of formalism, to capture concepts from the context information for helping to define data required by various service management operations. It also augments the expressiveness of the policy information model by adding domain-specific context data.

The diversity of languages used in management creates a corresponding diversity in management knowledge. However, semantic information can be managed by reasoners and semantic discovery tools that are capable of identifying the cognitive similarities between multiple concepts. As depicted in Fig. 4.1, if the Data Model A has some cognitive similarities with the Data Model B, the process to find such similarities is very complex, and in general fails when ontologies are not used due

J.M. Serrano Orozco, *Applied Ontology Engineering in Cloud Services, Networks and Management Systems*, DOI 10.1007/978-1-4614-2236-5_4,

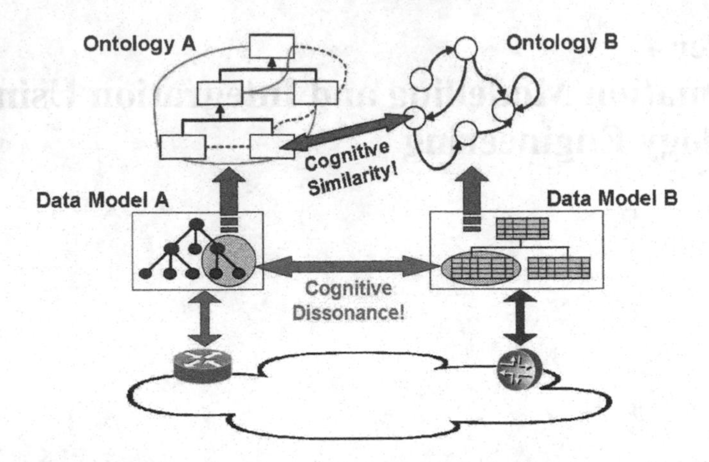

Fig. 4.1 Ontology integration process with other ontologies

to the lack of an underlying lexicon that enables different concepts to be semantically related to each other. However, when using ontologies, the terms in these two data models can be related to each other by using the formal linguistic relationships defined in the ontologies. In this chapter, the use of OWL as a formal language is explained and related to its capability to realize schema-based ontology matching and integration.

As a way of quick survey and as it was mentioned before, the use of standard languages, such as OWL, promotes the easy and flexible integration between ontologies and models. The characteristics of the ontology language define the clarity and quality of the knowledge that the ontology specifies. However, not all ontologies are built using the same set of tools, and a number of possible languages can be used. A popular language is Ontolingua [ONTOLINGUA], which provides an integrated environment to create and manage ontologies using KIF [Genesereth91]. Other languages, such as KL-ONE [Brackman85], CLASSIC [Borgida89] and LOOM [Swartout96], were defined according to domain-specific requirements.

Another standards-based approach is to follow the conventions defined in Open Knowledge Base Connectivity (OKBC) [OKBC] model, KIF or CL-Common Logic. Each of these languages are examples that have become the foundation of other ontology languages, and each specifies a language that enables semantics to be exchanged.

This chapter describes the ontology construction process. The phases for building an ontology are not detailed but the objective is to demonstrate the formal mechanism to represent information and most importantly the interactions between the different information domains. Formal concepts present in the information models are used and formally represented in this chapter. The relationships between the concepts from information models are then defined as part of the formalization process. This provides enhanced semantic descriptions for the concepts present in the information models.

The organization of this chapter is as follows. Section 4.2 provides a general description about the data and information model used as basis in our approach (i.e. the

Context Information) with the objective of clearly identifying the elements that exist in the information model and represent the objects definitions.

Section 4.3 introduces the policy model structures, first to define the structure of a policy and second to demonstrate the information links between different domains by using ontologies, that is Policy Information, and the Service Lifecycle Management Operations.

Section 4.4 provides the model interactions and formal representations, those information models are first studied over XML schemas to understand its components and then represented and integrated formally within the process for building an exemplar ontology; it means an ontology representation to understand the practical part of using ontologies for integration of data and information models. Specially, when each model is separately augmented semantically and then finally integrated by using ontologies.

Section 4.5 is related to the conclusions concerning this chapter.

4.2 Data and Information Modelling

Information technology advances, along with the evolution in communication services towards automated and mobile operations, demand the integration of information from heterogeneous, distributed technologies and systems. As described previously in this book, context-awareness plays an important role in next generation networks and communications systems.

Context-awareness is even more important as mobility becomes more feasible, since this enables uninterrupted services regardless of where the end user moves. Specifically, extensible context models enable the efficient representation for handling and distributing information and consequently, management systems that support pervasive services.

Interoperability of the information systems is required by complex scenarios, where diverse information models are involved in the interaction and exchange of information in order to support applications for service and network operations. When information interoperability is mentioned, scenarios with a mix of technologies supporting multiple services are considered; these scenarios have heterogeneous management systems and devices, and consequently different techniques and mechanisms for generating and sharing information.

In each one of these scenarios, the data models that the information systems use are different, which inhibits the sharing and reuse of network, user and context information. Hence, a way to achieve the efficient interaction between the systems is first, to create an information model that defines critical concepts in a technology neutral form, and second, to derive management models from this information model that support the free exchange of knowledge. This approach is explained in this section, and formalized by using ontologies to augment the information model in the subsequent sections.

The terms used in information models often constitute an informal agreement between the developers/programmers and the users of the information model (or even the data models that are derived from the information model). The language used to represent the data is usually informal. However, unless a single common information model (CIM) exists, there is no way to harmonize and integrate these diverse data models without a formal language, since informal languages may or may not be able to be unambiguously parsed. Hence, various initiatives have been proposed, with the objective to standardize and integrate such information.

These include information models such as the CIM [DMTF-CIM], authored by the Distributed Management Task Force (DMTF) and the Shared Information and Data model (SID) [TMF-SID], authored by the TeleManagement Forum (TMF). These two models are arguably driving the modelling task for the computer industry. However, both of these models represent vendor-independent data, as opposed to vendor-specific data, and hence are not adopted by many vendors. The documentation about CIM and SID information models can be found in [DMTF-DSP0201], [DMTF-DSP0005] and [TMF-SID], respectively.

However since the management perspective, those initiatives do not provide enough tools to integrate context information for management operations in communications systems (or vice versa: the modelling of commands issued by management systems that (re)configure devices and services). In addition, neither the CIM nor the SID provides any specific type of context model definition.

The level of formalism in both is compromised; the CIM does *not* use a standard language as UML (it has invented its own proprietary language), and while the SID is UML-based, other modelling efforts of the TMF are making compromises that are turning the information model into a set of data models. Indeed, the CIM is in reality a data model, as it is *not* technology-independent (it uses database concepts to represent its structure, which classifies it as a data model) [DMTF-CIM].

Thus, the real challenge is to promote information interoperability in heterogeneous systems combining network technologies, middleware and Internet facilities, to create an environment where the information between the devices and the applications and their services is always available. In this sense, the integration of information models is a non-trivial task, and it takes special care when diverse information models from different domains need to be integrated. Ontology engineering has been proven as a formal mechanism for solving problems in meaning and understanding; hence, ontology engineering appears to be as one of the best candidate for reconciling vendor- and technology-specific differences present in information and data models. While ontologies have been previously used for representing context information, currently most proposals for context representation ignore the importance of the relationships between the context data and communication networks.

The context information modelling activity, using ontologies, relies on knowing what elements of context are actually relevant for managing pervasive services; this in turn drives the selection and use of the proper ontology for discovering the meaning(s) and/or helping with reasoning about the context information in the networks [López03b]. In addition to this activity, within this process, context information has not been represented or considered yet in management ontologies as a relevant part for managing services. This section concentrates on describing the advantages of using ontologies to represent and integrate different types of context information from different information and data models into service management operations.

In the process of representing information from multiple data sources, many different mechanisms can be used. For example, XML [XML-RPC] has emerged as a widely accepted way of representing and exchanging structured information. XML allows the definition of multiple markup tags and constraints that can describe the relationship between information structures. In particular, an XML Schema (XSD) [XML-XSD] is a schema language, which means that it contains a set of grammatical rules which define how to build an XML document. In addition, an XSD can *validate* an XML document by ensuring that the information in the XML document adhered to a set of specific datatypes (i.e. it implies the existence of a data model that validates the content of the XML document). In this way, XSDs allow more control over the way XML documents are specified. Certain common datatypes are supported, and there is the ability to specify relationships and constraints between different elements of a document. However, an XSD can contain a complex list of rules, causing the XML document to be turned into a much more complex document for describing information. In addition to a non-user-friendly list of markup tags, XSDs define a limited set of datatypes, which can impede the natural representation of information.

The principal technical strengths of an XSD are that it has a text-based representation, which makes it easy to build tooling to process XSD documents. XSDs impose a strict syntax to permit the automated validation and processing of information in an unambiguous way, which is required by pervasive applications. The XSD/XML editor used is XMLSpy [XMLSPY].

The idea of creating ontologies to support the integration of diverse information models is the result of extensive research activity to find a solution to the problem, interoperability of the information, necessary for enabling integrated management. Integrated management is one of the most complex problems in ITC systems.

The use of ontologies is based on the assumption formal representations can enable computing systems to use any information that is relevant for a particular domain in other domain in mutual or individual benefit. This section is concentrated on this idea in order to improve and simplify the control of the various management operations required in the service lifecycle.

In this section, it is explained how to use information models to build a more formal ontology-based information models that structure, express and organize interoperable context information. The most important challenges for the integrated model is that the context information is dynamic (can change very quickly), and context information is naturally distributed across many layers in the systems. Thus, the models need to be much more robust, and at the same time, must be semantically rich and flexible to be used in multiple platforms and systems.

An ontology-based model considers the current status of the managed object, as well as current and future aspects of the context information describing the managed object, in order to determine if any actions are required (e.g. to govern the transitioning of the state of the managed object to a new state). This also applies to other managed objects that affect the state of the object being managed. Most current applications are narrowly adapted to specific uses, and do not provide sufficiently rich expressiveness to support such a generic context information model.

In the following sections, concept of using ontologies for context integration into service management operations is used, with a novel vision in which functional components and ontology-based middleware solution for context integration must contain. These results are presented as part of the construction process for the integrated model which ended up with an ontology-based model.

4.2.1 Context Information Model

A formal representation of context information is a complex task, the information is dynamic and there are many approaches to formalize the context information. In this section, a formal representation acting as base for representing context in the framework of the IST-CONTEXT project [IST-CONTEXT] is pointed and explained. In this sense, it is a necessity to develop extensible language that represents information to be shared. An extensible model which, each time a sub-system needs to access information, allows either for management operations or for configuring any application of personalized services and augment the information model with ontological data. In the described scenario, this section is centred as a practical example, that is IST-CONTEXT Project, both the user's context information as well as the information of the network can be used to customize services and resources controlling the deployment of new services.

Under the service described above, this section aims to describe how facts from diverse information and data models can be integrated with each other using a formal mechanism such as ontologies. This will enable applications to react to variations in context information, and adapt their functionality in an automated fashion. Such services are known as pervasive services, and defines the specific set of relevant context information (which can vary at any given moment) in order to determine which resources and services are offered to that user at that location at that particular time. This can be defined for new and/or existing functionality and services [Held02].

The complexity of designing and deploying pervasive services is very high; hence, appropriate tools and infrastructures are required to implement context-awareness as well as to acquire, share and reuse information. To achieve the goal of making a service context-aware in a formal way, the enrichment of the data models is crucial. For example, XSDs must specify the structure and integrity of the data in the form of sets or containers of information. Current practice just creates application- and technology-specific data models, which often creates management silos that inhibit reusability.

The most important challenge in pervasive services is the modelling and structure of context information [Brown00]. If the context can be formalized, then multiple applications can be controlled using the same information [Brown98]. The highly distributed nature of context information introduces the challenge of how a consistent and coherent context information model can be defined and managed.

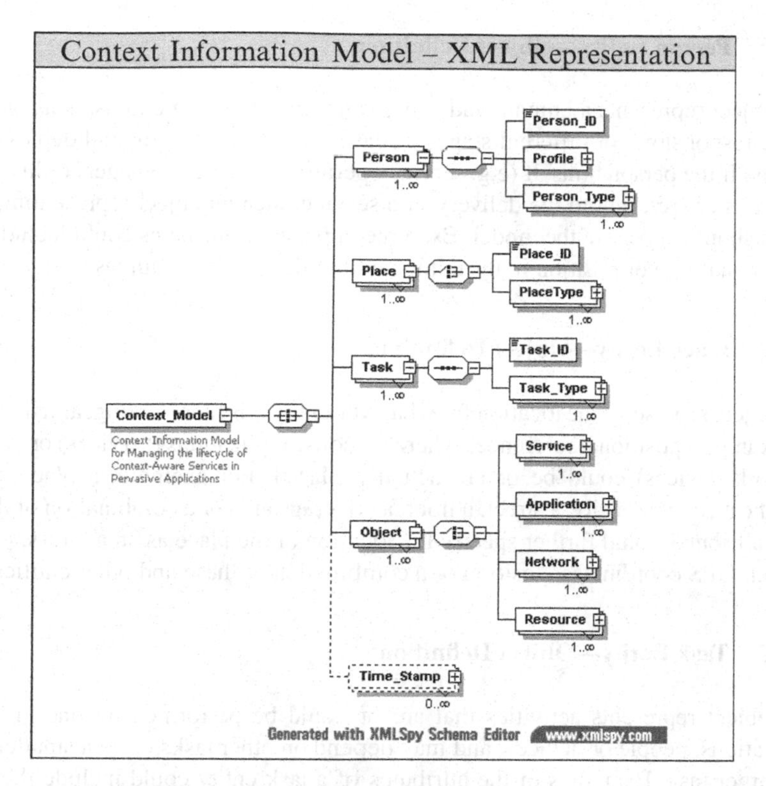

Fig. 4.2 Context information model in an XML representation

An additional challenge is the sharing and exchange of information between different levels of abstraction.

Figure 4.2 shows the context information model defined to capture and represent the explicit context information required to support pervasive services. The description of the context information model, with its set of general classes and sub-classes, is described in the following sub-sections.

4.2.2 *Definition of the Main Entity—Context Model Objects*

A simplified version of the context model is shown in Fig. 5.1. Descriptions about objects that are contained in this model are presented. The model has a small set of high-level classes of entities and relationships, in order to keep it conceptually simple. The model contains four main types of entities—person, place, task and object. These classes are defined by taking into account what they need to represent in the service provisioning process, and their relationships with various service lifecycle operations, and are described in the following sub-sections.

4.2.2.1 Person Entity—Object Definition

This object represents a human, and can be anyone, including end users as well as people responsible for different stages of the service provisioning and deployment process. If the person himself (e.g. service operator or service manager) or his characteristics are relevant to the delivery of a service, then an object representing that person should appear in the model. Example, a person's attributes could include the professional role or position in the company that this person occupies.

4.2.2.2 Place Entity—Object Definition

This object represents the location for whatever entity it is representing, and includes (for example) positional references where persons, applications (services) or objects (network devices) could be or are actually placed. For example, a place could describe a country, a city, a street, a floor, a coverage area or a combination of these. Basic attributes could further specify the location of the place as an address, phone number, GPS coordinates, position or a combination of these and other entities.

4.2.2.3 Task Entity—Object Definition

This object represents activities that are or could be performed by one or more applications, people or devices, and may depend on other tasks or be a smaller part of a larger task. Examples of the attributes of a task entity could include the start time, end time, due time and status (both in terms of success/failure as well as the percentage finished); clearly, additional attributes can be defined as well.

4.2.2.4 Object Entity—Object Definition

This object can represent any physical or virtual entity or device, such as a server, a router, a printer or an application. Examples of basic attributes of an object entity include the status (on, off, standby) of the entity and its description (technical or social). It is important to highlight a classification of context types must be used to help, examine and organize if there are additional pieces of context that can be useful in managing pervasive service applications.

4.3 Ontologies in Service and Network Management

4.3.1 Policy Structure

In this section, the policy representation, expressed in extensible markup language (XML) format, is described in detail, to enable their constituent parts to be better understood. This modelling process by using XML is considered as a reference in this book. There are four main policy models that are based on information models: (1) the IETF policy model, (2) the DMTF CIM, (3) the TMF SID and (4) the latest version of DEN-ng, being standardized in the ACF.

The IETF policy model is specified in RFC 3060 [IETF-RFC3060] and RFC 3460 [IETF-RFC3460]; the DMTF CIM is based on and extends these two models. Both the CIM and the IETF standard share the same basic approach, which specifies a set of conditions that, if TRUE, results in a set of actions being executed. In pseudo-code, this is:

IF a condition_clause evaluates to *TRUE*, subject to the evaluation strategy
THEN execute one or more actions, subject to the action execution strategy

In contrast, the SID, which is based on an old version of DEN-ng (version 3.5), adds the concept of an event to the policy rule. Hence, its semantics are:

WHEN an event_clause is received
IF a condition_clause evaluates to *TRUE*, subject to the evaluation strategy
THEN execute one or more actions, subject to the rule execution strategy

The latest version of DEN-ng (version 6.6.2) enhances this respecting the rules and adding alternative actions considered in a new pseudo-code, as explained in detail in [Strassner07a]:

WHEN an event_clause is received
IF a condition_clause evaluates to *TRUE*, subject to the evaluation strategy
THEN execute one or more actions, subject to the rule execution strategy
ELSE execute alternative actions, subject to the rule execution strategy

DEN-ng was still under revision at the time of book edition, but had some excellent improvements over the existing state-of-the-art. In this book, as part of the demonstrative example, a compromise between the existing IETF standards and the newly emerging DEN-ng architecture is therefore included.

The high-level description of policies follows the format:

WHEN an event_clause is received that triggers a condition_clause evaluation
IF a condition_clause evaluates to *TRUE*, subject to the evaluation strategy
THEN execute one or more actions, subject to the rule execution strategy
ELSE execute alternative one or more actions, subject to the rule strategy

The above pseudo-code makes an innovative compromise: it defines an event as a type of condition. The problem with the IETF standard is that no events whatsoever are mentioned. This means that there is no way to synchronize or even debug when a policy condition is evaluated, since there is no way to trigger the evaluation.

Hence, as part of the demonstrative example, which is a part of this section, the IETF model has been extended to add specific triggering semantics. This is indicated by the phrase “that triggers a condition_clause evaluation”; for reference, refer back to the SID pseudo-code, which only stated: “WHEN an event_clause is received”. The purpose of this addition is to explicitly indicate that an event triggers the evaluation of the condition clause.

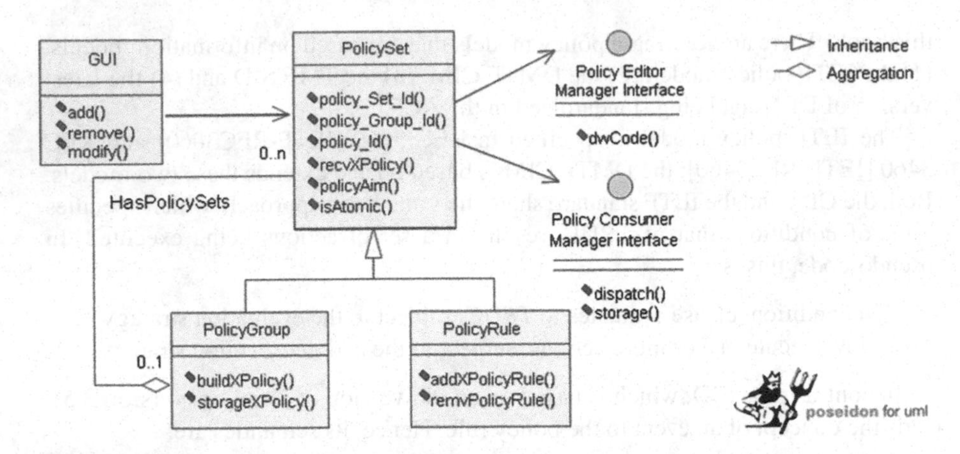

Fig. 4.3 The policy hierarchy pattern applied to the definition of PolicySets representation

The design improvement of the pseudo-code came after the work for the IST-CONTEXT architecture had been finished. Hence, to incorporate this work into the improved onto-CONTEXT architecture [Serrano07c], the model and ontology were redesigned to define events as types of conditions. In this way, the onto-CONTEXT architecture could remain compliant with the IETF standard but implement enhanced semantics.

The above description follows a simple syntax structure. Policies are not evaluated until an event that triggers their evaluation is processed. However, the rest of the policy uses implicit rules, which confers a poor semantic understanding. For this process, in order to enable the exchange of information contained in policies and the semantics describing what a policy might do and not do, and when it is executed, more semantic rules are mandatory.

4.3.2 Policy Hierarchy

The policies are structured hierarchically, in terms of Policy Sets, which can be either PolicyRules or PolicyGroups. The PolicyGroups can contain PolicyRules and/or other PolicyGroups. This is enabled through the use of the composite pattern for defining a PolicySet, and is shown in Fig. 4.3. That is, a PolicySet is defined as either a PolicyGroup or a PolicyRule. The aggregation HasPolicySets means that a PolicyGroup can contain zero or more PolicySets, which in turn means that a PolicyGroup can contain a PolicyGroup and/or a PolicyRule. In this way, hierarchies of PolicyGroups can be defined.

The order of execution of PolicyRules and PolicyGroups depends on the structure of the hierarchy (e.g. grouped and/or nested), and is controlled by its set of metadata attributes contained in Policy_Aim element. The service management policies control just the service lifecycle operations, and never the logic of the service. In this way, service management policies are used by the components of policy-based management systems to define the deployment of the service as a result of the management. For exemplification purposes, five types of policies covering

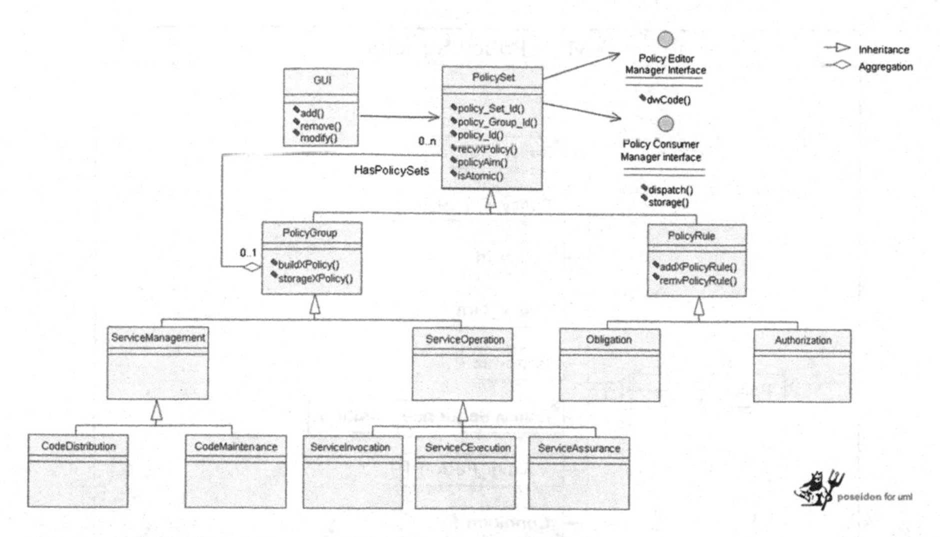

Fig. 4.4 Policy information model hierarchy representation

the service lifecycle have been defined; these policy types are structured around an information model whose most representative part is shown in Fig. 4.4.

4.3.3 Policy Model

Figure 4.5 shows the basic elements of a policy in XML format.

The structure of XML documents is dictated by the XSD against which that document is validated. Hence, the description of the main aspects of the XSD defined for Service Management Policies is included. Additionally, it is assumed that individual policies can be grouped into policy sets and policy groups. A policy set is an abstract concept that defines an intelligent container. Additionally in DEN-ng, a policy rule is actually an intelligent container [Strassner04], and is composed of events, conditions, actions and metadata. Individual policy rules can be grouped into a new structure, called a policy group, which in turn may contain other policy groups or single policy. A policy group adds additional metadata that can be used to cause all of its contained elements to be executed before or after other policies [Strassner04]. Hence, each individual policy will be identified using the vector<Policy_Set_Id, Policy_Group_Id, Policy_Id>. This hierarchical structure will be used to easily manage and organize the storage and processing of the policies loaded in the system.

4.3.3.1 Policy_Set_Id—Policy Set Identification Element

The Policy_Set_Id is the first of the identifier elements. The Policy_Set_Id element contains the identifier of the policy set to which the instance of this policy belongs.

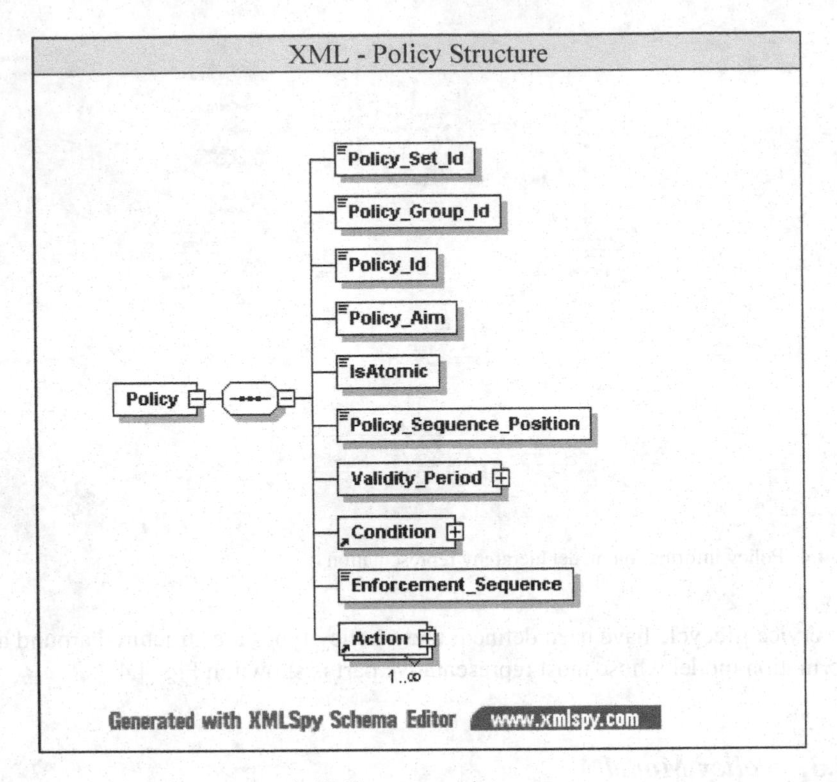

Fig. 4.5 Generic policy structure in XML representation

4.3.3.2 Policy_Group_Id—Policy Group Identification Element

The next field among the identifier elements is the Policy_Group_Id element. It contains the identifier of the group to which this instance of this policy belongs.

4.3.3.3 Policy_Id—Policy Identification Number Element

The third field among the identifier elements is the Policy_Id element. This element contains information to uniquely identify this policy instance from other instances of the same policy rule.

4.3.3.4 Policy_Aim—Policy Objective Element

This element defines information used to manage the policy when received from the Policy Definition System. This field specifies if this policy is a new policy to be loaded in the system, or if the policy identified by the above three-tuple (Policy_Set_Id, Policy_Group_Id, Policy_Id) has been already defined and, if so, has already been loaded into the system. In the latter case, this element defines whether this instance should replace the existing instance or not.

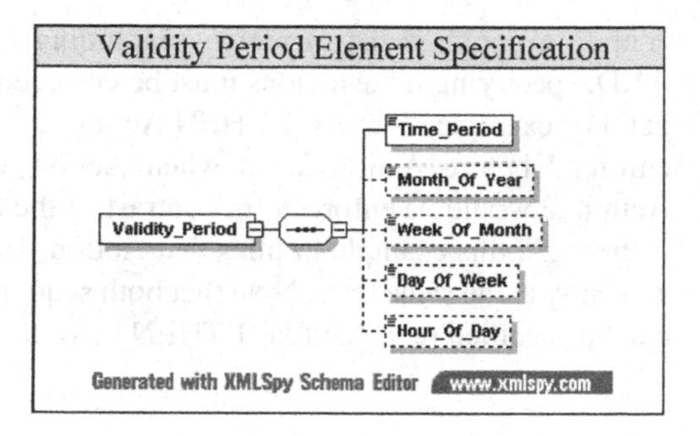

Fig. 4.6 Validity period element structure in XML representation

4.3.3.5 IsAtomic—Atomicity Definition Element

This field, dedicated to the management of this policy instance, defines the way to enforce the policy. The IsAtomic element is a Boolean value that defines whether concurrent execution is allowed. If concurrent execution is allowed, then multiple policies can be executed before their results are verified; otherwise, this policy must be enforced before starting the evaluation of the next policy in the sequence, that is the sequence in how the policy must be enforced cannot be interrupted.

4.3.3.6 Policy_Sequence_Position—Policy Sequence Element

This field, dedicated to the management of this policy instance, defines when this policy is evaluated with respect to other policy instances that are contained in this particular policy group.

4.3.3.7 Validity_Period—Validity Period Element

This element is used to express the policy expiration date. Usually, the expiration date is given as the time that the policy starts and finishes. Filters that specify further granularity can be also introduced. In Fig. 4.6, the structure of the Validity_Period element is shown. Note that the only mandatory element is the Time_Period element, which includes the start and stop times.

4.3.3.8 Condition—Condition Element

The Condition element includes all data objects, data requirements and evaluation parameters needed to specify and evaluate it. The Condition element contains three basic sub-elements: Condition_Object, Condition_Requirement and Evaluation_Parameters.

4.3.3.9 Enforcement_Sequence—Action Enforcement Sequence Element

The Enforcement_Sequence element specifies the time scheduling (sequential or concurrent) of the execution of a particular set of policy actions. This element must contain an ordered list of actions, including logical "connectors" like THEN or AND, specifying if the actions must be enforced in a sequential or concurrent manner. For example, "Action_1 THEN Action_2" would imply that Action_1 must be enforced before Action_2, so when Action_1 was successfully enforced, then Action_2 would be enforced. In contrast, if the action specified as "Action_1 AND Action_2", this example implies that Action_1 and Action_2 can be enforced concurrently at the same time. Note that both sequential as well as concurrent execution can be specified (e.g. "action_1 THEN (action_2 AND action_3)").

4.3.3.10 Action Element

The Action elements contain all the information needed for the enforcement of the specific action. The Action element contains three basic sub-elements: Action_Parameter, Enforcer_Component, Enforcement_TimeOut and Success_Output_Parameters. The structure of the Action element is shown in Fig. 5.6.

4.4 Ontology Engineering and Model Interactions

Multiple ontology development publications describe different methodologies for creating domain- and application-specific ontologies [Gruber93b], [McGuiness02], [Neches91]. In this chapter, the use of these methodologies to create an ontology is applied to the domain of pervasive services and its associated management operations.

While the methodologies mentioned have been used to develop, build and verify the consistency of the ontology, most ontology-based proposals do not address the relationship between the different domains of context data, policy models and communication networks and its application and usage as studied in [Serrano07c]. This section describes the guidelines for a novel approach, whose specific aim is to integrate ontological data with information and data models, thereby creating a powerful semantic representation of formal knowledge oriented for supporting pervasive service and network management operations. While there is no correct way to model a domain, let alone build an ontology, there must be a set of common, tested steps that can address ontology development activity in order to share and reuse concepts [FIPA-SC00094].

In business support systems, applications and services are usually organized into separate administrative domains that are often independent of each other. This results in a large diversity of terms and relationships between the terms used in each administrative domain, which in turn adversely affects the ability to share and reuse information between these domains.

This section proposes the definition and integration of classes from the three different domains of context data, policy models and communication networks. The following sub-sections review these domains: first, context integration, where the modelling and applicability of the information is the main goal of using ontologies to support the integration of the context information to help define the context-awareness of the services provided; second, the policy model, where the ontologies help represent the terms, elements and components of a policy and the relationships between them, and are also used to establish the relationships or links with the context information; third, communications networks, where the ontologies represent the elements, operations and components that manage the lifecycle operations of pervasive services.

The following section describes these three domains, along with the associated class definitions. It is important to highlight the interaction between classes from different domains, and how they interact with each other.

4.4.1 Context Information Model Based on Ontologies

This section does not propose a new information model for context, although it does extending existing definitions in earlier works and compiled in the state-of-the-art in Chap. 3 (Sect. 3.3.1.1 for more details). The model used is composed by the four concepts, such as person, place, task and object, which have been found to be the most fundamental data required for representing and capturing the notion of context information.

The information requirements and the revision of the state-of-the-art can be reviewed in [Bauer03], [Debaty01], [Eisenhauer01], [Gray01], [Henricksen04], [Korpiää03b], [Schmidt02], [Starner98] and other research works.

Ontologies are used to formalize the enhancements made to the extension of the context model, in this section the interaction between domains is specified and represented. This reflects one of the objectives of this book, which is to model the entities defined in the information model using a formal language based on ontologies, and to represent the context information. Specifically, in this section UML class diagrams are used to define basic context and management information, and then enhances this information to create the formal ontology.

The context representation using UML classes consists of the definition of classes and their relationships. Relationships are defined as a list of elements that have a relationship, such as a dependency or an aggregation, to another set of elements. The Context class can be related to a set of classes that represent the specific context information that can be shared.

Figure 4.7 shows the context information model upper level ontology. The context representation is structured as a set of abstract classes describing a physical or virtual (i.e. logical) object in the service domain. Attributes and relationships can be optionally specified to further define the characteristics of and interaction between different aspects of the context. The Context class is related to the Object, Person, User, Place, and Task classes. Each of these are implemented as class containers.

This enables the definition of each to be inherently extensible; it also enables the information in each of the components that is placed in a class container to be

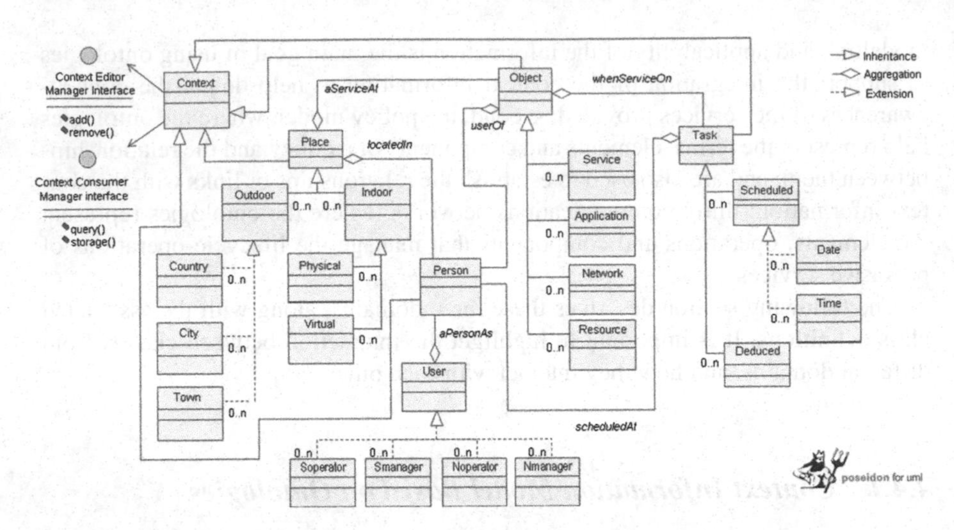

Fig. 4.7 Context information model class diagram

defined as aggregations and/or compositions to reflect additional semantics. For example, the User class can be instantiated as up to four different types of users (i.e. Soperator, Smanager, Noperator and "Nmanager") where each user type is involved in a particular type of service management operations.

The relationships between context classes and classes from other domains are described by the set of UML models and snippets of the ontology and its interactions, as shown below and in the following sub-sections.

To implement this extended context information model, and to adequately capture its associated semantics, ontology-based vocabularies for expressing concepts describing context information are used. Note that if UML models are used, the level of expressiveness is limited to the keywords or specific types of information defined in the model. Ontology concepts facilitate the association of related information, since pattern matching as well as linguistic relationships (e.g. synonyms, holonyms, meronyms, hypernyms and hyponyms) can be defined and associated with each other to represent context, network and service operations as well as the management operations required to manage them.

Figure 4.8 shows the ontology interactions between the classes of the information model and the context information. Note that the structure merely suggests certain vocabularies related to each other for creating associations between concepts. For example, a "User" is a "Person" who isDefinedAs a "ContextEntity" and isStoredAs part of the "DataModel" with some other specific properties.

By using ontologies, the nature of information for context data can be defined in a more formalized manner. Furthermore, the data can be *managed* formally, and is made extensible through the ability to refine any of its model elements (e.g. classes, attributes and relationships). This facilitates the sharing and reusing of information. In Fig. 4.8, the isDefinedAs, anActivityFor, isUsedFor and isLocatedAt relationships define Person, Task, Object and Location class instances as being related to a "Context Entity". The isStoredAs relationship means that a ContextEntity can be

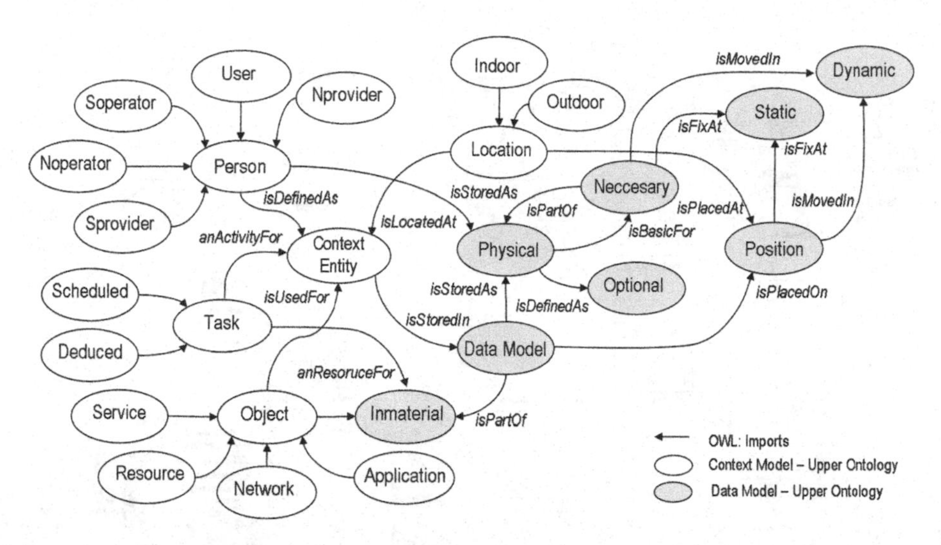

Fig. 4.8 Context information model ontology interactions map

stored as a "Data Model". These five relationships collectively constitute the bridge to formalize the concept of context. Similarly, the relationships between Person and the Physical, Place, Position, Task and Object concepts with the Inmaterial concept define these as semantic relationships.

For example, assume that a pervasive context-aware service is being defined. When an end user appears in a specific WiFi coverage area, the first operation that is required is to determine how those data are related to the context of that end user. Hence, these data need to be related to a "Context Entity" that includes that end user.

Using the ontology representation shown in Fig. 4.8, it can be seen the "Context Entity" is related to a "Data Model", which is in turn related to the location of the end user, both through the isLocatedAt relationship as well as the *isPlacedOn* relationship. This latter relationship associates the position of the end user with the context, which means that static as well as dynamic locations are automatically accounted for. Note the difference between these relationships and, for example different types of Locations (i.e. indoor or outdoor)—the former require instance data (and hence are defined only in the Data Model) while the latter can be statically defined, since "indoor" and "outdoor" concepts are predefined. This means that when the Context Entity is being used to trigger the service, the location of the user can be used to do so.

4.4.2 Policy Information Model Based on Ontologies

The second domain is the policy-based management domain. Chapter 2 reviews the state-of-the-art of policy information models where the ontology-based policy approaches have been studied (please refer to Sect. 3.4.1.2 for more details).

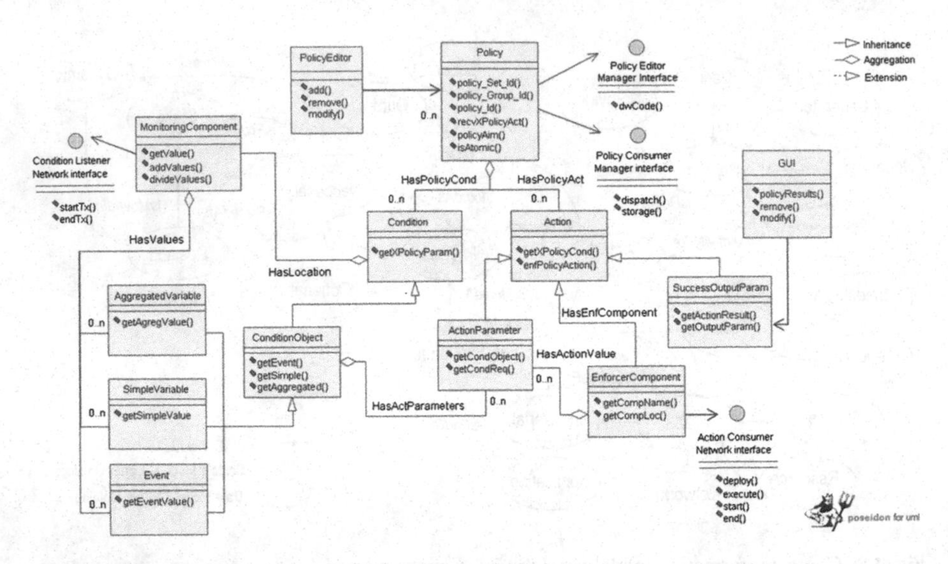

Fig. 4.9 Policy information model class diagram

However, from that proposals study [Damianou02], [Sloman99c], [Uszok04], [Kagal03] and others in [Guerrero07], no one has used ontologies before to capture the knowledge concepts from context information and integrate it with concepts that policies contain; thus as result of this, integration could help at service and network management systems to support management operations controlling the service lifecycle.

Figure 4.9 shows the policy information model class diagram. The main policy class components that are technology independent are the Condition, Action, MonitoringComponent, EnforcerComponent, PolicyEditor and ConditionObject; the rest of the components are required to realize the policy-based paradigm. The Condition and Action components are the main components that implement the semantics of a policy. The Condition component is related to the MonitoringComponent to check the state and performance of the managed entity that the policy governs, while the EnforcerComponent is used to ensure that the operations defined by the actions of a policy were successfully executed.

The ConditionObject refines the generic concept of a policy condition for use in pervasive service applications. It defines three main types of variables (i.e. Aggregate, Simple and Events) that can be used to form the condition of the policy which is to be evaluated. Event defines significant occurrences that are used to trigger the evaluation of the condition to determine the proper set of actions (if any) that should be executed in response to this event.

The SimpleVariable class represents contextual and management data that can be monitored using an appropriate MonitoringComponent, while the AggregatedVariable is a set of Events, SimpleVariables and even AggregatedVariables that use arithmetic operations to derive results that can in turn be evaluated.

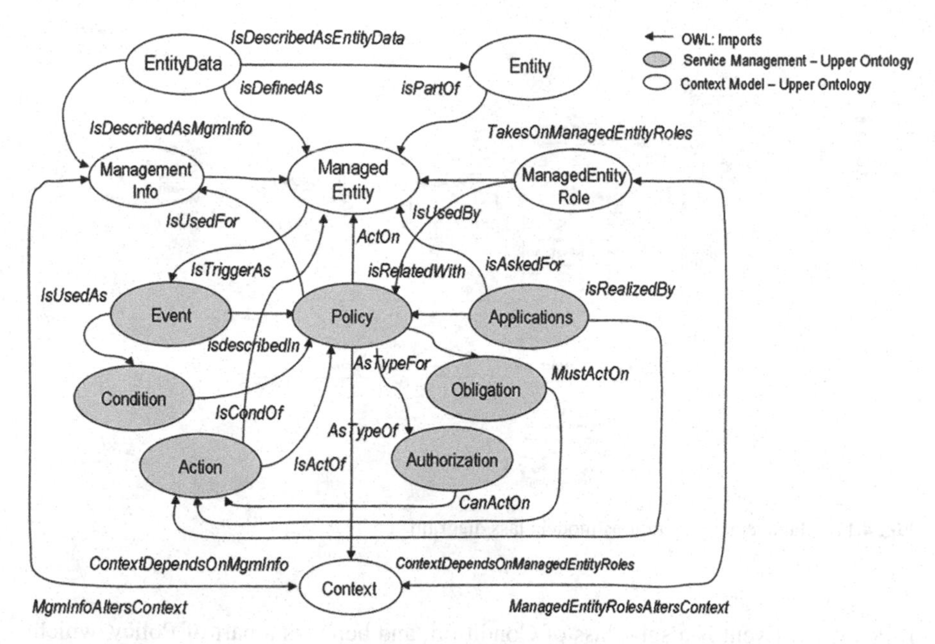

Fig. 4.10 Policy information model ontology interactions map

The SuceesOutputParam class defines parameters to verify if the result of the actions of a policy has been successfully executed. The Policy Editor represents an application that can create, read, modify and/or delete Condition and/or Action components from a policy. This policy information model is intended to support management lifecycle operations, as it emphasizes the structure of the policy as well as the components use to monitor and enforce the policy.

The next step in the process to formalize this information model is to use abstract class structures to represent its main concepts. This enables us to focus on the interactions between the different constituent classes of the model in order to build the associated ontology. The ontology is structured as a set of classes describing objects in the policy information model domain; the attributes and relationships of the information model are used to build properties and relationships in the ontology. Figure 4.10 shows how the different concepts that represent classes from the policy information model interact with other abstract concepts in the ontology.

This type of diagram is used in order to create more powerful semantic expressions that express how different concepts interact. In the specific case of the represented ontology, these interactions are used to express how the context information is related to different management operations. For example, using the same typical example of a user appearing in a WiFi coverage area whose context-aware functionality is governed by policies, the Event can be thought of as "user arrives in WiFi area".

This is the first part of the context information that is constructed when this event is received. In this example, the Event is a type of Condition that is used to trigger the reading (and possible operating on) management data. This is represented as

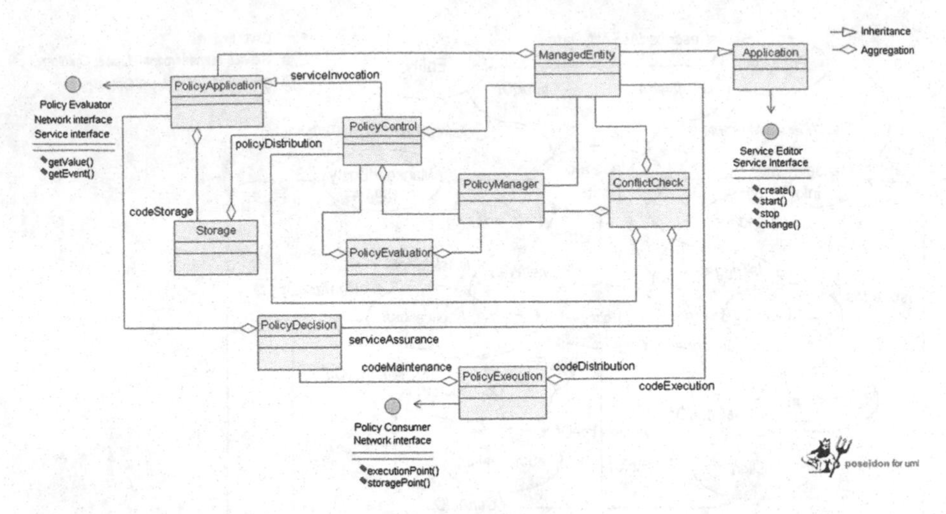

Fig. 4.11 Management operations model class diagram

follows. An Event is a sub-class of Condition, and hence is a part of Policy, which is a sub-class of ManagedEntity.

The ManagementInfo class represents different types of EntityData that contains management and network information that is used to manage and monitor ManagedEntities; hence, ManagementInfo helps further to define the particular Context. The Event triggers the evaluation of a policy Condition. If the condition is true, appropriate actions are executed. In this example, the Condition can be thought of as an instance of the "isAskedFor" relationship. The Service is represented by one or more ManagedEntityRoles which abstract the functionality that is to be delivered to the user. Variations on this service or its functionality directly affect the Context, which enables closed loop management to be performed when the user arrives to the coverage WiFi area.

4.4.3 Operations Management Model Based on Ontologies

The third domain is the set of management operations for services management. This domain has been defined in the framework of business-oriented technologies, but to date has never been related to either network operations or especially to management applications. The proposal in this section is hence novel, since context information is used to control management operations of pervasive services.

Figure 4.11 shows a set of interactions between various policy-based management operations. This class diagram depicts some of the important service lifecycle operations that must be controlled by the integrated context information. Here, integration explicitly means the definition of relationships in the information model and the formalization of those relationships by the ontology. The operations are structured as relationships between basic service management components (abstract classes with attributes) in a service management system.

The service management operations are the result of policy tasks that execute in response to the evaluation of certain conditions related with the service lifecycle. For example, a policy could request certain information from a service listener and, based on the data received, distribute and/or execute one or more sets of service code.

The information to be delivered by the listener, in this example, can then be a value that makes the policy evaluation true (i.e. equal to its expected value), which then results in executing one or more actions. In the class diagram of Fig. 4.11, the evaluation of the value, coming from a context variable, is managed by the *PolicyApplication,* and such information triggers a *serviceInvocation*. One of these policy actions can be the *policyDistribution* to certain "Storage" points. The PolicyDecision decides what actions must be executed and then pass control to PolicyExecution.

A *serviceInvocation* can be signalled by a ManagedEntity containing the context values to be evaluated by the PolicyApplication, or from the Service Listener. The PolicyExecution is responsible for the distribution of service code and service policies as well as their deployment (as a result of *codeDistribution* and the codeMaintenance operations).

The PolicyEvaluation helps the PolicyManager to make decisions based on the values of the relevant context information, which can be measured or computed. Finally, the ConflictCheck is responsible for ensuring that the set of current policies do not conflict in any way (e.g. the conditions of two or more policies are simultaneously satisfied, but the actions of these policies do different operations to the same ManagedEntity).

Figure 4.12 shows the service management components and the associated management operations involved in the service lifecycle process. Those operations are typically represented as policy management concepts when UML class diagrams are being used. However, this is a novel vision using ontology class interaction maps to integrate information between different domains, as it is shown this is an important interaction, and it is explained shortly as follows.

The novel use of this interaction map enables the visualization of the semantic relationships necessary when different classes are being related to each other. This is especially important when these classes are from different domains. One example is the ability to view semantic descriptions of the context data to identify the source of the information. For example, this can be used to determine if the data is being produced from end users (e.g. personal profile, service description or variables) or network devices (e.g. server properties or traffic engineering values).

The main operations that a policy-based system can execute on a ManagedEntity are shown in the Ontology class interactions map, where the PolicyManager works to satisfy the *policyConditionsRequestedBy* relationship between the PolicyManager and the PolicyApplication.

The PolicyApplication is related to the ManagedEntity in several ways, including directing which management and/or context information is required at any given time.

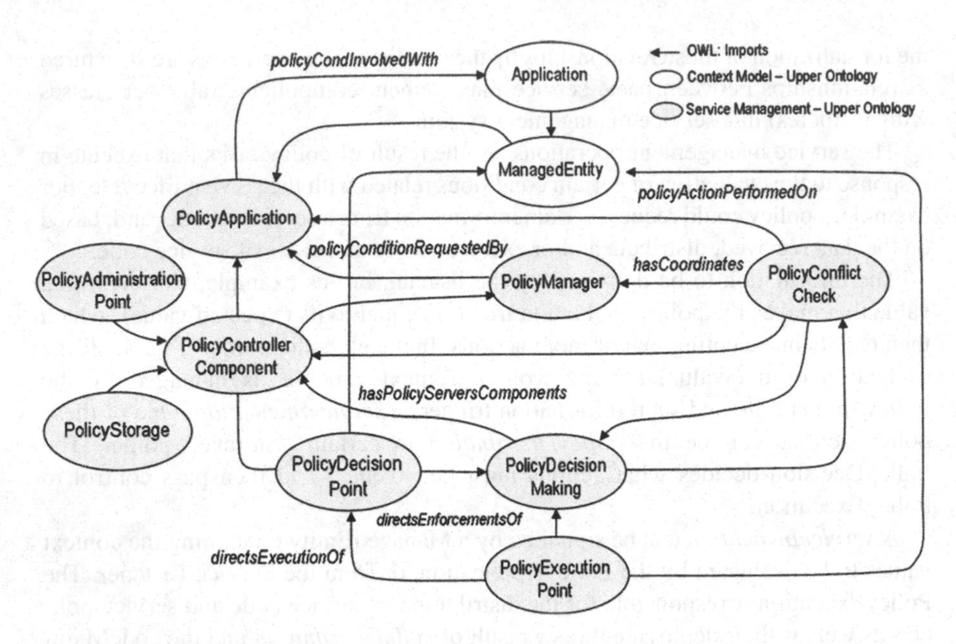

Fig. 4.12 Management operations model ontology interactions map

Since a ManagedEntity is directly related to different Applications, those Applications can act as brokers or wrappers for service provisioning. In addition, some information from the Application can be used by policies. The PolicyManager governs and coordinates monitoring as well as management decisions (and their enforcement) using various appropriate entities, such as PolicyExecutionPoints and PolicyDecisionPoints. The PolicyApplication controls all policies in the system related with the Application to share and reuse information contained in the policies.

An important aspect of policy-based service management is the deployment of services in multiple network elements. For instance, when a service will be deployed over any type of network, decisions must be taken in order to determine which network elements will support the service and thus have their configurations changed.

This is most effectively done through the use of policies that map the user and his or her preferences to the capabilities of the set of network elements that will support the service. Moreover, policies can also control service invocation and execution, which enables a flexible approach for customizing one or more service templates to multiple users. Furthermore, the maintenance of the code realizing the service, as well as the assurance of the service, can all be driven by policy management [Sloman94b], [Strassner04]. Finally, when the system senses undesirable service behaviour, one or more policies can define what actions need to be taken to solve (or at least mitigate) the problem.

4.4.4 Service Lifecycle Control Model Based on Ontologies

To date, initiatives have used policy management approaches to tackle the problem of fast and customizable service delivery. These include OPES [OPES] and E-Services [Piccinelli01]. In this book, the use of the policy-based management paradigm to control the full service lifecycle using ontologies is a step further in this area, and it is presented as a novel example to do data integration by using formal mechanism as ontologies. In addition, to complement the use of ontology to formally represent information and integrate data, a functional architecture which takes into account the variation in context information and relates those variations to changes in the service operation and performance for services control is described.

The managing concepts for service lifecycle operations are contained in the ontology for integrated management. The integration of concepts from context information models and policy-based management constitutes the foundations of the semantic framework, which is based on the construction of a novel ontology model for service management operations. A demonstration of the concepts of this approach is one of the most important contributions in the area of knowledge engineering and telecommunications [Serrano07a].

Policy-based management is best expressed using a restricted form of a natural language than a technical or highly specialized language that uses domain- and technology-specific terms for a particular knowledge area. A language is the preferred way to express instructions and share data, since it provides a formal rigor (through its syntax and grammatical rules) that governs what makes up proper input. An ideal language is both human- and machine-readable, which enables systems to automate the control of management operations. However, in network management, multiple constituencies are involved (e.g. business people, architects, programmers, technicians and more) and all must work together to manage various aspects of the system. Each one of these constituencies has different backgrounds, knowledge and skill sets; hence, they are represented using different abstraction levels.

The language and the interactions within these different abstraction levels are shown in Fig. 4.13. While most of these constituencies would like to use some form of restricted natural language, this desire becomes much more important for the business and end users (it even becomes undesirable for some constituencies, such as programmers, that are used in formal programming languages). This notion was described previously as the Policy Continuum in [Strassner04], where each constituency is assigned a dialect of the language to use.

The introduced and depicted ontology is a global domain ontology that captures the consensual knowledge about context information, and includes a vocabulary of terms with a precise and formal specification of their associated meanings that can be used in heterogeneous information systems. The ontology was designed to enable policy-based management operations to more easily share and reuse data and commands. The basic approach is to use the Policy Continuum to connect a service creation view to a service execution view using multiple dialects of a common language.

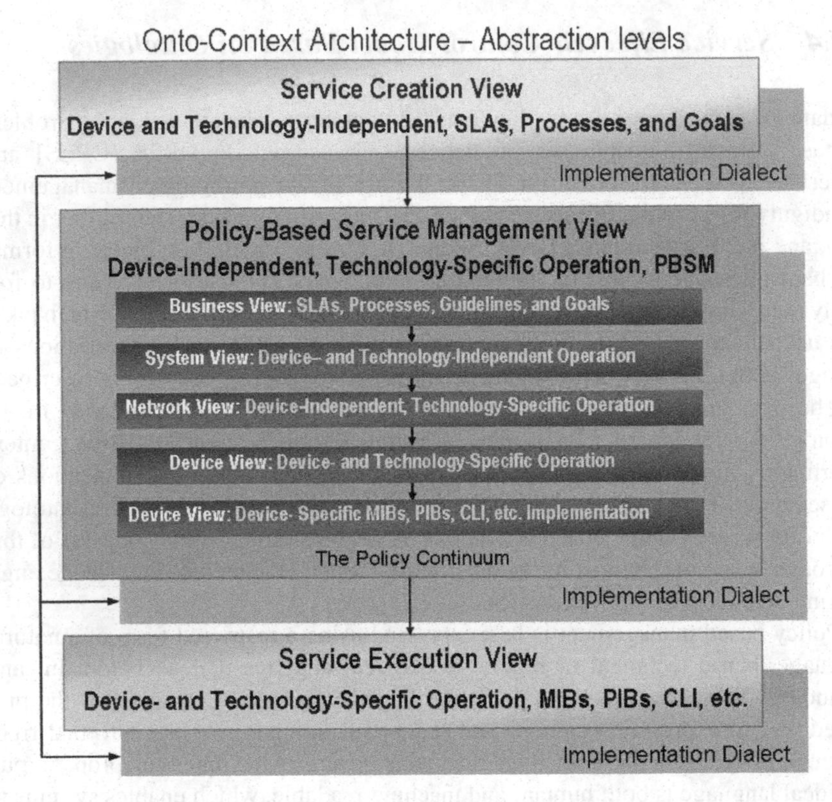

Fig. 4.13 Policy-based management languages and Policy Continuum representation

This ensures that the different needs and requirements of each view are accommodated. For example, there is a distinct difference between languages used to express *creating* services from languages used to express the *execution* and *monitoring* of services. This also enables this approach to accommodate previously incompatible languages, due to their different structures, formats, protocols or other proprietary features. While there are some existing policy languages that have been designed (e.g. Ilog Rules [ILOGRULES] and Ponder [Damianou01]), each of these (and others) use different commands, which have different semantics, for executing the same instruction when a policy is being evaluated.

The approach illustrated in this section uses a set of ontologies that capture the syntax and semantics from different areas, but at the same time provide a level of constituency by mapping those terms and phrases to the same expressive language. This is done by using a set of languages based on OWL to ensure platform independence. Since OWL is based on W3C standards [W3C], this approach takes advantage of a popular existing standard, and hence makes it more appealing for adoption. By augmenting this with formal graphical representations using UML [OMG-UML] and describing the OWL syntax using XSDs, this approach takes advantage of the large variety of off-the-shelf tools and freely available software providing powerful and cost-effective editing and processing capabilities.

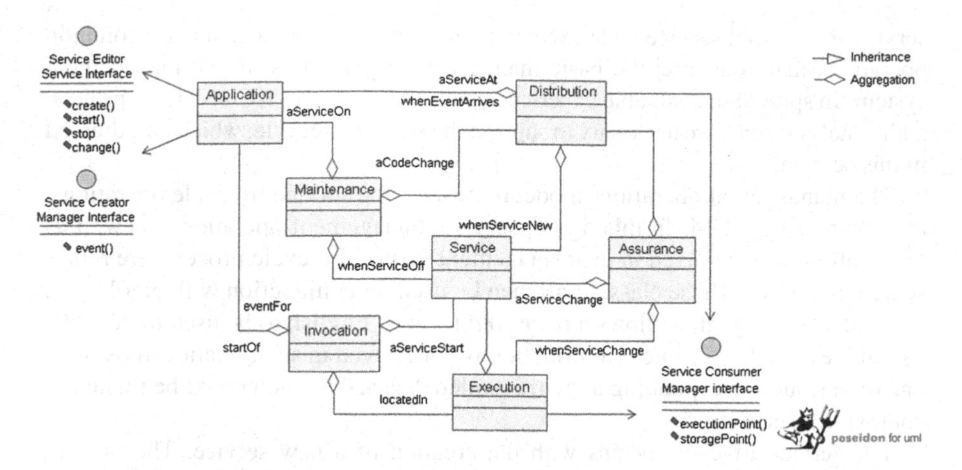

Fig. 4.14 Service lifecycle management interactions representation

Each one of the implementation dialects shown in Fig. 4.14 is derived by successively derivated vocabulary and grammar from the full policy language to make a particular dialect suitable to each particular level of the Policy Continuum. For some levels, vocabulary substitution using ontologies enable more intuitive GUIs to be built.

This ontology modelling work is based on the definition of a policy information model that uses an ontology to help match a set of dialects to each different constituency. This follows the methodology described in the Policy Continuum. Specifically, a context model and a policy model are combined to define management operations for pervasive services in an overall framework, which provides a formal mechanism for management and information systems to share and reuse knowledge to create a truly integrated management approach. This exemplar and instructive work is aligned with the activity of DEN-ng and it constitutes a future research line for converging with such model to build up a DENON-ng and ontology for the DEN-ng information model.

The ontology is driven by a set of pervasive service management use cases that each requires a policy-based management architecture as represented in this section. The ontology is founded on using information models for context information and policy management to promote an approach to integrated management, which is required by both pervasive as well as autonomic applications. The combination of context-awareness, ontologies and policy-driven services motivates the definition of a new, extensible and scalable semantic framework for the integration of these three diverse sources of knowledge to realize a scalable autonomic networking platform.

Policies are used to manage various aspects of the service lifecycle. Thus, the scope of this example addresses the various service management operations identified from the research and experience acquired by the active participation in the IST-AUTOI [AUTOI] and IST-CONTEXT Project [IST-CONTEXT]. These service management operations include code distribution, service code maintenance, service invocation, service code execution and service assurance, and are common operations that constitute the basic management capabilities of any management system. In spite of the valuable experience gained from the IST-CONTEXT project, additional activities are necessary to support the service lifecycle, which is addressed in this section.

The management operations model represents the service lifecycle operations, as shown in Fig. 4.14. In this figure, service management operations, as well as the relationships involved in the management service lifecycle process, are represented as classes. These classes will then be used, in conjunction with ontologies, to build a language that allows a *restricted* form of English to be used to describe its policies. To do so context information is underlayed in such relationships, with one or various corresponding activities called "*events*", which could be related to context information.

The service lifecycle begins with the creation of a new service. The Service Editor Service Interface acts as the application that creates the new service. Assume that the service for deploying and updating the service code in certain network nodes has been created. This results in the creation of an event named "*aServiceOn*", which instantiates a relationship between the Application and Maintenance classes. This in turn causes the appropriate policies and service code to be distributed via the Distribution class as defined by the "*aServiceAt*" aggregation. The service distribution phase finds the nearest and/or most appropriate servers or nodes to store the service code and policies, and then deploys them when the task associated with the "*eventFor*" aggregation is instantiated. When a service invocation arrives, as signalled in the form of one or more application events, the invocation phase detects these events as indication of a context variation, and then instantiates the service by instantiating the association "*aServiceStart*". The next phase to be performed is the execution of the service. Any location-specific parameters are defined by the "*locatedIn*" aggregation. The execution phase implies the deployment of service code, as well as the possible evaluation of new policies to monitor and manage the newly instantiated service.

Monitoring is done using the service consumer manager interface, as it is the result of associations with execution. If maintenance operations are required, then these operations are performed using the appropriate applications, as defined by the "*aServiceOn*" aggregation, and completed when the set of events corresponding to the association "whenServiceOff" is received. Any changes required to the service code and/or polices for controlling the service lifecycle are defined by the events that are associated with the "*whenServiceNew*" and "*aServiceChange*" associations.

The service management operations are related to each other, and provide the necessary infrastructure to guarantee the monitoring and management of the services over time. The UML design shown in Fig. 4.14 concisely captures these relationships, thus the pervasive service provisioning and deployment is on certain manner assured to provide service code and policies supporting such services to the service consumers.

4.5 Conclusions

In this chapter …

The main advantage of using an ontology-based approach to integrate and represent context information and management operations has been introduced and discussed. The isolation achieved between explicit context and implicit context information enables the interaction and exchange of information to be done by the management systems when context information is being used as information for managing purposes.

Ontologies have been used for specifying and defining the properties that represent the context information and the management operations that support pervasive services in the communications networks. It has been depicted and demonstrated that ontologies used for representing the context information and the management operations for pervasive services enable services to exhibit an adaptive behaviour.

Ontology engineering for supporting the management of services through context integration enables context-awareness to be better implemented by enabling the operation of management systems to use context information and policy-based management mechanisms. The use of ontology engineering is versatile and solves some of the problems in the area of pervasive service management.

The ontology used as example is an ontology-based information model for pervasive applications that integrates its information model describing different aspects of context with ontologies that augment the model with semantic information; these augmented data enable context-aware service requirements to be modelled. Structuring this as a formal language using OWL provides extensible and efficient context information for managing services.

According to the ontology representation, pervasive services can be automatically customized, delivered and managed. The ontology is suitable not only for representing and defining context information concepts but also for defining management operations that motivate and promote the integrated management of pervasive services.

The contributions of this chapter are focused on defining the basis for functional middleware technologies, defining an extensible ontology for the robust support and management of pervasive services. Future work will continue the study of formal ontologies and their interaction with information and data models. One of the key areas that are addressed is how to automatically validate, as much as possible, the relevance and correctness of both information and data models as well as appropriate ontologies in pervasive systems.

Chapter 5
Ontology Engineering as Modelling Process in Service Management Architectures

5.1 Introduction

Management of next generation networks (NGNs) and next generation services is affected by the Internet and its new business models demands. As clear examples of this change are web services [W3C-WebServices] and most recently the emerging cloud services. As a result, new operations, service descriptions and management functions supporting communications systems are necessaries alike the organization of data seeking information interoperability. This chapter presents the conceptual mechanisms for representing and governing the various aspects of the service lifecycle introducing autonomic management principles. These aspects affect both the organizational view of the service lifecycle and its operational behaviour, and use semantic control in each to achieve interoperability in the information necessary to control and manage pervasive services. This chapter refers to using semantic information to govern the management operations for controlling the entire service lifecycle using ontologies to provide self-management operations.

This chapter is dedicated to describe modelling processing outcomes as applied to pervasive services management operations. The outcomes are result of the necessity for representing, in a formal way, the service lifecycle and service management operations for context-aware applications. This chapter introduces a formal representation which pointed towards realistic applications in autonomic communications and cloud computing environments. However, those formalisms can be applied to other areas where services management processes work in a similar fashion. This chapter also introduces the concept of using semantic control rules for supporting the task of defining semantic interactions when using ontologies based on the formal representation for the service lifecycle introduced and discussed.

The organization of this chapter is as follows. Section 5.2 provides a general scope for introducing the service lifecycle from the perspective of pervasive service management. It provides a brief description of the service lifecycle phases. Section 5.3 presents the study of the management operations for pervasive services, and its interactions are identified with the objective of facilitating the control of the

J.M. Serrano Orozco, *Applied Ontology Engineering in Cloud Services, Networks and Management Systems*, DOI 10.1007/978-1-4614-2236-5_5,

service lifecycle phases. Section 5.4 introduces examples of semantic-based rules and proposes that using semantic control, as provided by ontology-based functions for defining the semantic interactions between ontology-based information models, service control task are simplified. The ontology-based functions are used then for the control of service operations and to, for example, control services in the cloud or other services that require linked data or information exchange.

In this chapter, service management operations are described and represented with the combination of ontologies and information models. Through the study of formal ontologies for the integration of information and the functional validation in pervasive systems, this chapter aims to highlight the importance of using ontologies and information models, and their integration, towards the support of autonomic communications and cloud computing.

5.2 The Service Lifecycle—A Management Perspective

In the context of managing communication systems that continually are increasing in complexity, the integration of information controlling network and services management constitutes a real challenge. It is highly desirable that the information and operands are accessible from their corresponding information sources, and that they can be distributed across the network to satisfy service and user requirements in a cross-layered environment in a consistent, coherent and formal way.

In this section, the use of policies for service management is explained, and the policies are augmented with the use of formal ontologies [Serrano08]. This enables the management systems to support the same management data to accommodate the needs of different management applications through the use of rich service control semantics, some of which are being specified in the ACF. Service management applications for pervasive systems highlight the importance of formal information models in policy-driven service architectures. The policies are used in managing various aspects of the lifecycle of services. It is important to identify what is meant by the term "service lifecycle" in the framework of this book. Thus, the following sections describe the pervasive service lifecycle as a set of stages, including its complete customization, management, deployment, execution and maintenance.

Currently, the TMF is specifying many of the management operations in networks for supporting services [TMN-M3050], [TMN-M3060], in a manner similar to how the W3C specifies web services [W3C-WebServices]. However, a growing trend is to manage the convergence between networks and services (i.e. the ability to manage the different service requirements of data, voice and multimedia serviced by the same network), as well as the resulting converged services themselves. The management of NGN pervasive services involves self-management capabilities for improving performance and achieving the interoperability necessary to support current and next generation of Internet and the now emerging exponentially cloud services.

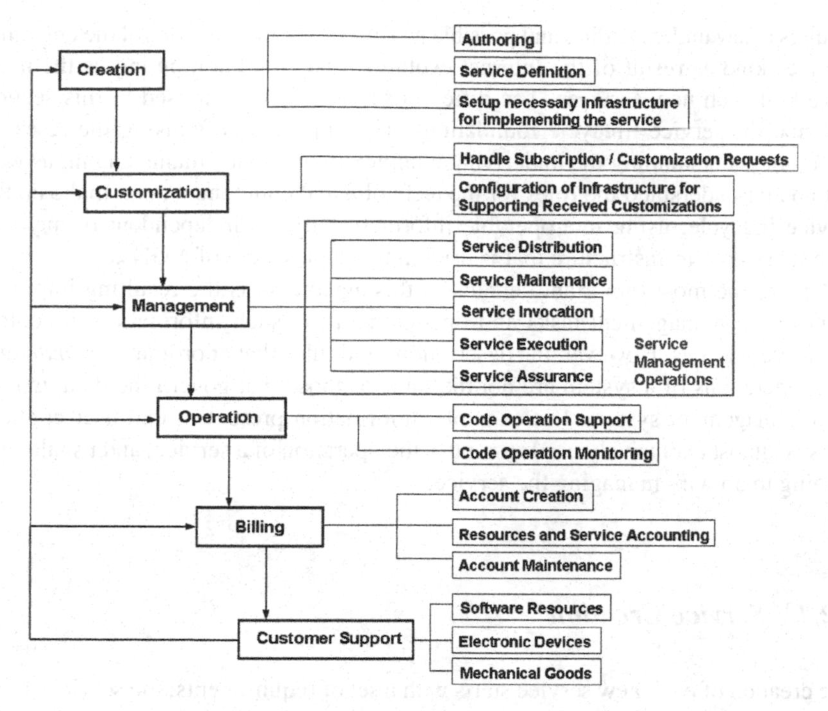

Fig. 5.1 Pervasive service lifecycle representation

Self-management features depend on both the requirements and the capabilities of the middleware frameworks or platforms for managing information describing the services as well as information supporting the delivery and maintenance of the services. The representation of information impacts the design of novel syntax and semantic tools for achieving the interoperability necessary when NGN resources and services are being managed. Middleware capabilities influence the performance of the information systems, their impact on the design of new services and the adaptation of existing applications to represent and disseminate the information.

The vision of self-management creates an environment that hides the underlying complexity of the management operations, and instead provides a façade that is appealing to both administrators and end-users alike. It is based on consensual agreements between different systems (e.g. management systems and information support systems), and it requires a certain degree of cooperation between the systems to enable interoperable data exchange.

The organizational view for the pervasive services lifecycle, which can be divided into six distinct phases with specific tasks is depicted in Fig. 5.1.

Management operations are the core part of the service lifecycle, and are where the contributions of this chapter are focused. The management phase is highlighted in Fig. 4.1. Creation and customization of services, accounting, billing and customer support are not described as they are broadly studied; however, management requires a new understanding and possible new ways and tools to control the emerging services kind as result of the Internet evolution and cloud computing as the main drivers of such new services. The different service phases exposed in this section describe the service lifecycle foundations. The objective is focusing the research efforts in understanding the underlying complexity of service management, as well as to better understand the roles for the technological components that make up the service lifecycle, using interoperable information that is independent of any one specific type of infrastructure that is used in the deployment of NGNs.

One of the most important benefits of this agreement is the resulting improvement of the management tasks and operations using such information to control pervasive services. However, the descriptions and rules that coordinate the *management* operations of a system are not the same as those that govern the data *used* in each management system. For example, information present in end-user applications is almost exclusively used to control the operation of a service, and usually has nothing to do with managing the service.

5.2.1 Service Creation

The creation of each new service starts with a set of requirements; the service at that time exists only as an idea [Serrano08]. This idea of the service originates from the requirements produced by market analysis and other business information. At this time, technology-specific resources are not considered in the creation of a service. However, the infrastructure for provisioning this service must be abstracted in order to implement the business-facing aspects of the service as specified in a service definition process.

The idea of a service must be translated into a technical description of a new service, encompassing all the necessary functionality for fulfilling the requirements of that service. A service is conceptualized as the instructions or set of instructions to provide the necessary mechanism to provide the service itself and called service logic or most commonly typified as SLO's.

5.2.2 Service Customization

Service customization, which is also called authoring, is necessary to enable the service provider to offer its consumers the ability to customize aspects of their services according to their personal needs and/or desires. Today, this is a growing trend in web-services, cloud computing and business orientation designs. An inherent portion of the customization phase is an extensible infrastructure, which must be able to handle service subscription/customization requests from administrators as well as consumers.

5.2.3 *Service Management*

The main service management tasks are service distribution, service maintenance, service invocation, service execution and service assurance. These tasks are described in Sect. 5.3.

5.2.4 *Service Operation*

The operation of a deployed service is based on monitoring aspects of the network(s) that support the service, and variables that can modify the features and/or perceived status of the communications. Usually, monitoring tasks are done using agents, as they are extensible, can accommodate a wide variety of information, and are easy to deploy. The information is processed by the agent and/or by middleware that can translate raw data into a form having explicit semantics that suit the needs of different applications.

5.2.5 *Service Billing*

Service billing is just as important as service management, since without the ability to bill for delivered services, the organization providing those services cannot make money. Service billing is often based on using one or more accounting mechanisms that charge the customer based on the resources used in the network. In the billing phase, the information required varies during the business lifecycle, and may require additional resources to support the billing. The make up of different billing infrastructures is out of scope of this chapter. However, since the management of service information is just as important as the maintenance of the service, information and processes defined here can be used for both processes.

5.2.6 *Customer Support*

Customer support provides assistance with purchased services, computational resources or software, or other support goods. Therefore, a range of services and resources related are required to facilitate the maintenance and operation of the services, and additional context (and sometimes the uncovering of implicit semantics) is necessary in order for user or operators to understand problems with purchased services and resources. A basic business requirement is that the capability for an easy and rapid introduction of new services is mandatory.

5.3 Service Management Operations and Semantic-Based Services Control

In this section, the management operations of a pervasive service and its interactions are identified as distinct management operations from the rest of the service lifecycle phases. Figure 5.2 depicts management operations as part of the management phase in a pervasive service lifecycle.

In this book, the idea of context information is explained, and other types of knowledge are required in order to provide better management services. An enhanced understanding of the semantics of a set of service management operations to enable current and future pervasive service management is crucial to achieve enhance semantic management control. Further explanations in this section drive to posit that increased semantic control is best realized through the use of ontologies, which can integrate context information with other management information to achieve interoperability in the various management systems used by pervasive systems.

Management systems must be capable of supporting semantic variations of the information and react to those variations in application-specific ways (e.g. to enable context-awareness). Management systems must also be flexible enough to manage current as well as future aspects of services variations in response to variations in context information.

An important aspect of policy-based service management is the deployment of services using programmable elements. For instance, when a service is going to be deployed, decisions have to be taken in order to determine which network elements are going to be used to support the service. This activity is most effectively done through the use of policies that map the user and his or her desired context to the capabilities of the set of networks that are going to support the service. Moreover, service invocation and execution can also be controlled by policies, which enable a flexible approach for customizing one or more service templates to multiple users.

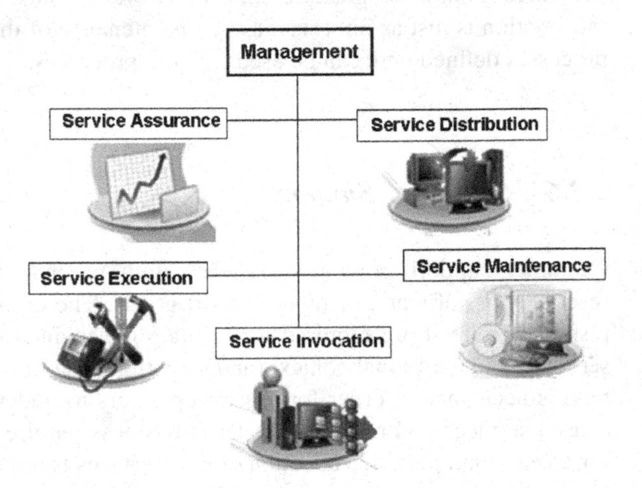

Fig. 5.2 Pervasive service management ant the most common operations

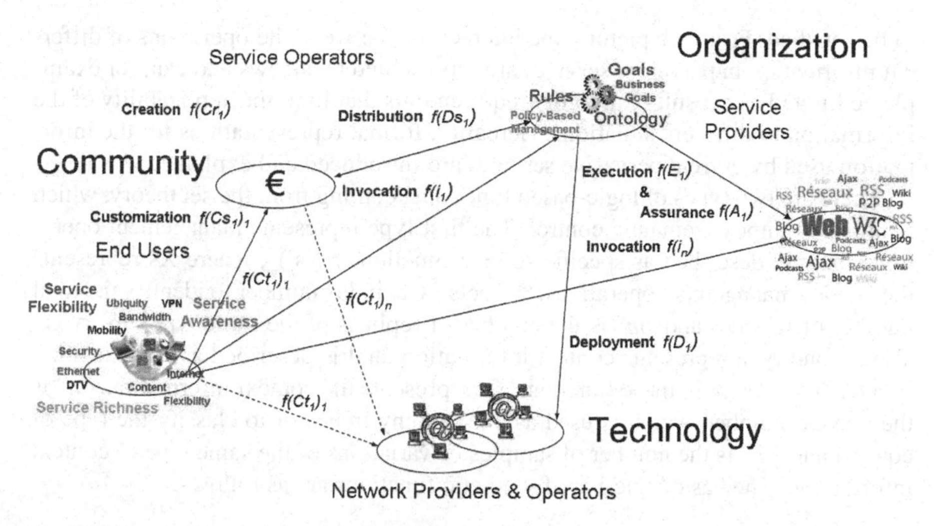

Fig. 5.3 Organizational view of information interoperability interactions

Policy management is an effective mechanism for maintaining code to realize the service, changes in the service and assurance of the service. For example, when variations in the delivery of the service are sensed by the system, one or more policies can define the set of actions that need to be taken to solve the problem. In this way, the use of policies enables different behaviour to be orchestrated as a first step to implement self-management.

Policy-based management systems use a combination of abstractions (usually via data and/or information models) and logic-based rules (usually via ontologies) to support the activities and actions of basic management operations independent of the reasoning engines or mechanisms they implement. Policy-based management offers the ability to adapt the behaviour of the management systems through a consistent interface that is inherently extensible.

Important feature is the extensibility to translate the variations in behaviour to associated variations in the communication infrastructure via a set of policy rules. The adaptation is implemented by ontologies, which represent the different semantics as a set of keywords for managing the systems or to trigger the execution of appropriate policy rules. In other words, particular descriptions and/or concepts can be used to generate semantic control to enable management systems to react to changes in context information.

Semantic rules provide new capabilities to management systems for processing information and reacting to semantic variations of context and management data [Ghidini01]. If it is assumed that context information can help in the control operations of the service lifecycle, then a number of questions arise. However, while the information can be transparently shared between different management systems of different organizations, there is an inherent limitation in sharing information.

As illustrated in Fig. 5.3, the problem of different organizational nodes attempting to exchange information or pieces of information generates the scenario need to be satisfied. From the picture, the interactions between the operations of different information in pervasive services are represented by arrows and can, for example be limited as a result of mobility requirements that limit the applicability of the information to different situations. Hereafter, formal representations for the information used by current pervasive services are introduced and explained.

There are two types of logic-based functions, coming from the set theory, which is in use to impose semantic control. The first type represents management operations that are described as specific service functions $f(Xs_n)_m$, where Xs represents the service management operations, "*n*" acts as an index number to identify the total number of services and "*m*" is the number of replicas of the same type of service. The second type represents context information, and is described as context functions $f(Ct_n)_m$, where in these functions, Ct represents the context information, "*n*" is the context number, which is used as a taxonomy indicator to classify the type of context and "*m*" is the number of samples or variations of the same type of context information. The basic logic-based semantic functions are as follows:

$$f(Xs_n)_m \rightarrow \text{service functions} \quad \text{where } n > 1 \text{ and } m > 0, \tag{5.1}$$

$$f(Ct_n)_m \rightarrow \text{context functions} \quad \text{where } n > 0 \text{ and } m > 0. \tag{5.2}$$

The functions are representations for expressions that identify variations in the semantic values of the information, and can be operated on using logic or mathematical expressions. Logic and mathematical operations are used for both business-oriented solutions and management information from possibly heterogeneous networks that could be distributed in different databases (e.g. from different sensors).

The functions are independent of each other. The way to relate context information to certain services using such information is by generating inclusive functions "*F*" which contains sets of service and context functions. In these kind of functions, the constraints "$n>1$" and "$m>0$" are forced to be rewritten as "$n \geq 1$" and "$m \geq 1$", since the service and at least 1 sample of context must both exist. The expression representing such conditions is as follows:

$$F[f\{(Ct_n)_m\}f\{(Xs_n)_m\}] \rightarrow \text{context functions related with service functions,} \tag{5.3}$$

where $n \geq 1$ and $m \geq 1$.

Integrating the expression by set theory arguments and using summation:

$$\sum F[\{(Ct_n)_m\}\{(Xs_n)_m\}] \rightarrow \text{summation of context functions with service functions,} \tag{5.4}$$

where $n \geq 1$ and $m \geq 1$.

The proposed service lifecycle management operations do not assume a "static" information model (i.e. a particular, well-defined vocabulary that does not change) for expressing policies. This enables multiple policies to be processed, since context conditions can be defined dynamically (e.g. new variable classes can be defined at run-time, so those variables acting as context information are dynamic).

The expression representing such dynamic context variations when multiple "*p*" service policies are being executed is as follows:

$$\sum_{\mathrm{Xs}=1} F[\{(\mathrm{Ct}_n)_m\}\{(\mathrm{Xs}_n)_m\}]p \rightarrow \text{context function referring to multiple ``}p\text{'' services,} \quad (5.5)$$

where $n \geq 1$ and $m \geq 1$ and $p \geq 1$ (when at least one service is active).

The essence of this function is to associate context information functions with service functions. This is accomplished using a set of policies.

A set of policies is represented by the notation "*pn*". Also, the initial condition of management systems requires an initial set of policies represented as "*ps*". Specifically, the context functions can either match pre-defined functions that manipulate existing schema elements, or can be used to extend these schema elements to represent new information that is required by the service.

The final representation of the function is:

$$\sum_{\mathrm{Xs}=1}^{ps+pn} F[\{(\mathrm{Ct}_n)_m\}\{(\mathrm{Xs}_n)_m\}]p^{ps+pn} \rightarrow \text{context function referring to multiple ``}pn\text{'' policy services,} \quad (5.6)$$

where $n \geq 1$, $m \geq 1$, $p \geq 1$, Xs is valid for $\mathrm{Xs} \geq 1$ to "$ps+pn$" (the total number of service policies).

The total number of semantic variables, plus the number of non-repeated relationships between the context data and the service management operations data, is required in order to solve this semantic-based logic function. The work to formalize the information by using first order logic models is studied in [Katsiri05] and others. For example, assume that one service has five policies "$pn=5$" and as initial condition "$ps=1$," those policies are used to represent its service management operations $\mathrm{Xs} \geq 1$ to $\mathrm{Xs}=ps+pn$, and furthermore assume that no context variations exist. This can be represented as:

$$\sum_{\mathrm{Xs}=1}^{ps+pn} F[\{(\mathrm{Ct}_n)_m\}\{(\mathrm{Xs}_n)_m\}]1^{(ps+pn)} \rightarrow \text{semantic interactions or functions using schema elements,} \quad (5.7)$$

where $ps=1$, $pn=5$ and $P=1$ and this conditions are valid for $\mathrm{Xs}=1$ to $ps+pn=6$.

Representing the expression when $Ct_1 = 1$ (no context variations) the function obtained is:

$$\sum_{Xs=1}^{ps+pn} F[\{(Ct_1)_1\}\{(Xs_1)_1\}]1^{(ps+pn)} \rightarrow \text{semantic interactions or functions with no context variations,} \tag{5.8}$$

where p is valid for $Xs = 1$ to $ps + pn = 6$.

Developing the expression when $Ct_1 = 1$ (no context variations) the function is:

$$\begin{aligned}\sum_{Xs=1}^{ps+pn} F[\{(Ct_1)_1\}\{(Xs_1)_1\}]1^{(ps+pn)} &= F[\{(1_1)_1\}\{(1_1)_1\}]1^6 + F[\{(1_1)_1\}\{(2_1)_1\}]1^6 \\ &+ F[\{(1_1)_1\}\{(3_1)_1\}]1^6 + F[\{(1_1)_1\}\{(4_1)_1\}]1^6 \\ &+ F[\{(1_1)_1\}\{(5_1)_1\}]1^6 + F[\{(1_1)_1\}\{(6_1)_1\}]1^6 \\ &= 21 \end{aligned} \tag{5.9}$$

Finally, giving values to the function the total of operations is obtained:

$$\sum_{Xs=1}^{ps+pn} F[\{(Ct_1)_1\}\{(Xs_1)_1\}]1^{(ps+pn)} = 21 \quad \text{(total of semantic interactions or functions).} \tag{5.10}$$

In this example, the number of schema elements required is 21. To prove if this approximation is correct, the use of graphs for representing the schema elements in the above example is presented in next sections. At the same time, those graphs help to depict the service management operations and their interactions. This feature itself is a requirement of the design for pervasive systems in order to achieve rapid context-aware service introduction and automated provisioning, and is supported by our approach. Note that the schema elements of the ontology can themselves serve as a guide to know if the number of semantic interactions is correct.

5.3.1 Interactions in Service Management Operations

To support the interactions between organizations in a pervasive service, a logic allowing decision-making and the subsequent choosing of alternatives as result must be used. The use of policy-based paradigm in conjunction with ontologies allows this, policies are used to control which parts of the ontology are managed, and ontologies can be used to represent the different semantics that are operated on using logic functions. Policies by themselves usually have no logic operations, but policy-based management using ontologies allows this semantic control.

Use cases are an excellent way to define the interaction between stakeholders in the system, and can be used to simplify and better understand the activity in

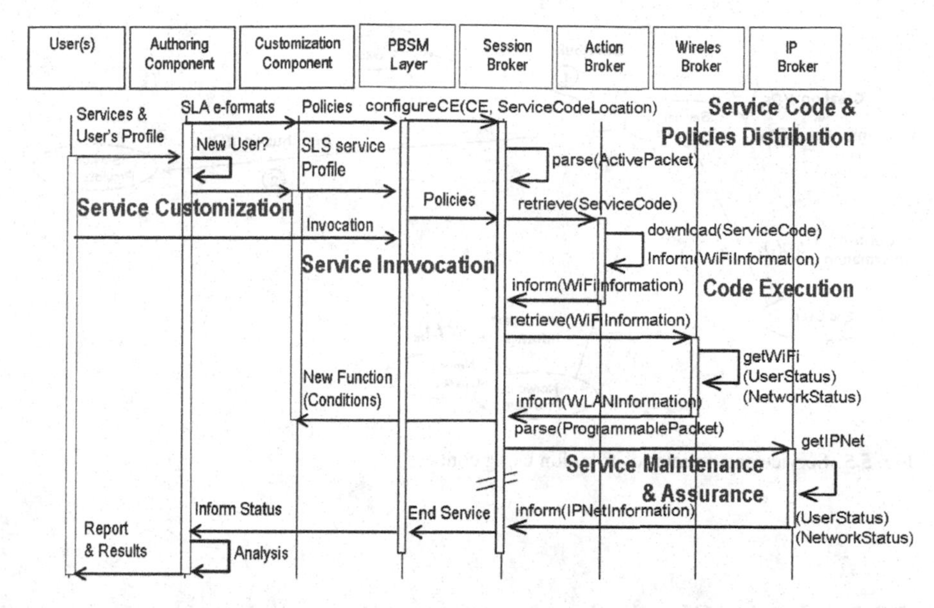

Fig. 5.4 Representation for service management operations—sequence diagram

complex scenarios. Semantic functions are enabled using the concepts that build up later the ontologies, as explained in the beginning of this section (i.e. Sect. 4.3). By now in this section, both UML use cases as well as other approaches are used to extract the meaning of the use case(s) and links, and hence provide meaning to the system representations. The basic management operations to support the lifecycle of pervasive services will be described.

Another tool for representing service management operations that is able to identify the associated semantic interactions is the use of sequence diagrams to represent the variations and conditions in the service [Kouadri04], as is shown in Fig. 5.4. This shows a sequence diagram for pervasive services, where the common operations required to manage pervasive services are described as a communication process between two actors in the service lifecycle. The basic operations are highlighted as common operations present in most pervasive services as a result of the research activity and analysis of the scenarios examined (see Chap. 6 for details).

Requirements dictated by pervasive services in terms of information [Park04] use different types of models that have been proposed in order to enhance information interoperability [Park04], [Fileto03]. However, none of these requirements have been related with service functions or management operations. Knowing the information requirements dictated by pervasive services and the different types of models that have been proposed so far for information interoperability, a relationship between those models and management functions must be established. The purpose of this relationship is to quantify how much each relationship supports or contributes to controlling the service lifecycle phases supporting the management operations.

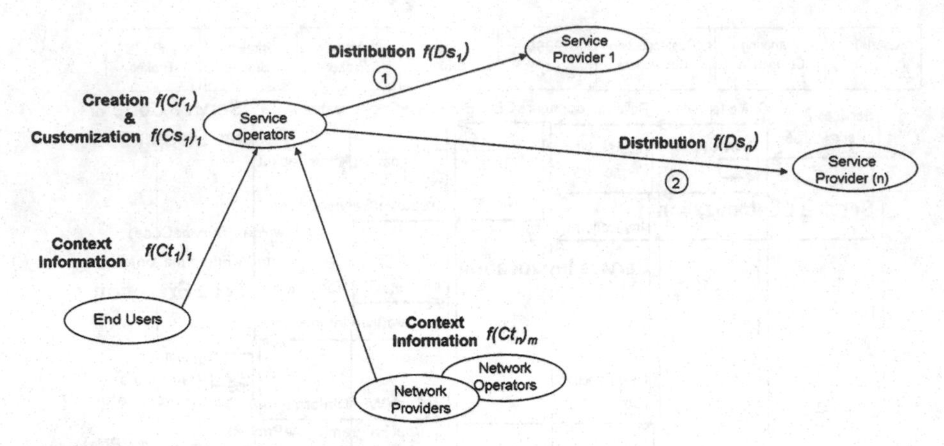

Fig. 5.5 Service and policies distribution using context

5.3.2 *Service Distribution*

This step takes place immediately after the service creation and customization in the service lifecycle. It consists of storing the service code in specific storage points. Policies controlling this phase are termed code distribution policies (*Distribution*). The mechanism controlling the code distribution will determine the specific set of storage points that the code should be stored in. The enforcement will be carried out by the components that are typically called Code Distribution Action Consumers. A high level example of this type of policy is as follows:

"*If* (customized service event $f(Cs_1)_1$ is received)
then (distribute service code $f(Ds_1)$ in optimum storage points selection with parameters $f(Ds_n)$"

Figure 5.5 represents how context information, as represented by the event in a function $f(\mathrm{Ct}_1)_1$, triggers the distribution of the code and the policies in functions $f(\mathrm{Ds}_1)$ through $f(\mathrm{Ds}_n)$ as a result of context variations $f(\mathrm{Ct}_n)_m$, where n is the index number to identify the type of context and m is the number of samples or variations of the same type of context information.

This figure shows two service functions $f(\mathrm{Ds}_1)$ and $f(\mathrm{Ds}_n)$, as well as two context functions $f(\mathrm{Ct}_1)_1$ and $f(\mathrm{Ct}_n)_m$. In this example, it is assumed no context variations and only one type of information; thus, in this graph, can be observed that, according to (5.6), two ((1)(2) = 2) schema elements are required to represent the two semantic interactions that are necessary to in this ontology representation and that must be considered when the ontology is being created.

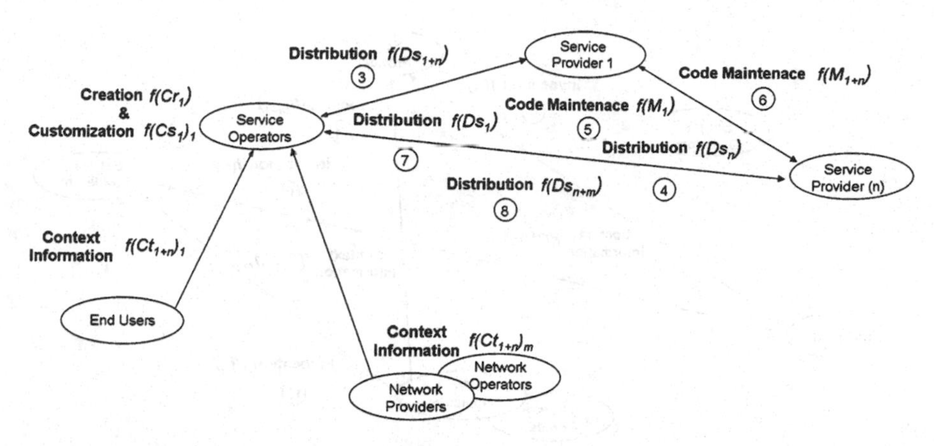

Fig. 5.6 Service maintenance as a result of context variations

5.3.3 *Service Maintenance*

Once the code is distributed, it must be maintained in order to support updates and new versions. For this task, special policies have been used, termed code maintenance policies (*CMaintenance*). These policies control the maintenance activities carried out by the system on the code of specific services. A typical trigger for these policies could be the creation of a new code version or the usage of a service by the consumer. The actions include code removal, update and redistribution. These policies will be enforced by the component that is typically named the Code Distribution Action Consumer. Three high level examples of this type of policies are shown here:

"*If* (new version of service code defined by $f(Ds_{1+n})$ is TRUE)
then (remove old code version of service $f(Ds_1)$ & (distribute new service code, function of $f(M_{1+n})$)"

"*If* (customized service code expiration date defined by $f(Ct_{1+n})_m$ has been reached)
then (deactivate execution for service *f(Ds1+n)* & (remove code of service, in function of *f(M1+n)*)"

"*If* (The invocation's number for service is defined by the function $f(Ct_{1+n})_m$)
then (distribute more service code replicas $f(Ds_{1+n})$ to new Storage Points as defined by the function $f(M_{1+n})$)"

Figure 5.6 represents how context information $f(Ct_n)_m$ controls the maintenance of the code and the policies in functions $f(M_{1+n})$ as a result of context variations $f(Ct_n)_m$, triggering the deployment of $f(Ds_1)$ through $f(Ds_{1+n})$, which in turn invoke new services. In these functions, *n* is the index number that identifies the type of context, while *m* is the number of samples or variations of the same type of context information.

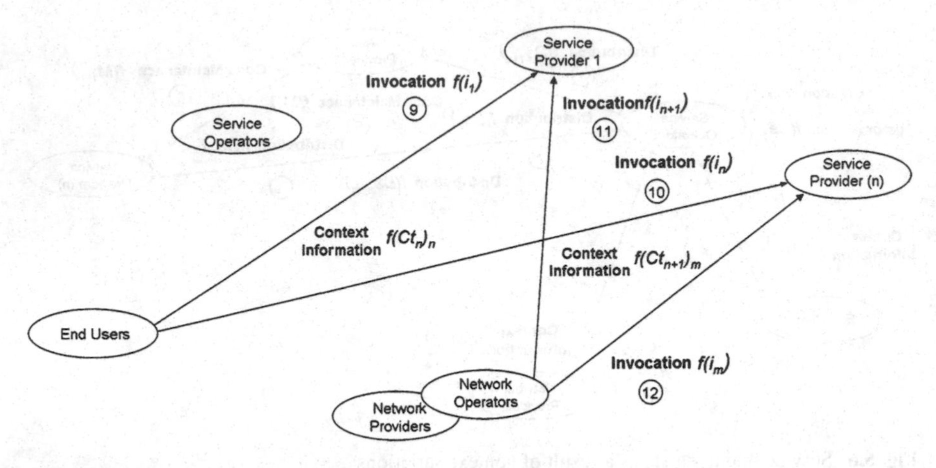

Fig. 5.7 Context information triggering the service invocation

This figure presents six service functions $f(\mathrm{Ds}_1)$, $f(\mathrm{Ds}_n)$, $f(\mathrm{M}_1)$, $f(\mathrm{Ds}_{1+n})$, $f(\mathrm{Ds}_{n+m})$ and $f(\mathrm{M}_{1+n})$, *the latter three functions arising from the* effect of the code maintenance operation. Two context functions, $f(\mathrm{Ct}_1)_1$ and $f(\mathrm{Ct}_n)_m$, are also defined. Since it was already assumed that no context variations are present, and that only one type of information exists, then in this graph it is observed (again using (5.6)) that six schema elements ((1)(6)=6) are required. This is equivalent to requiring six semantic interactions that must be present in the ontology representation.

5.3.4 Service Invocation

The service invocation is controlled by special policies that are called *SInvocation* Policies. The service invocation tasks are realized by components named Condition Evaluators, which detect specific triggers produced by the service consumers. These triggers also contain the necessary information that policies require in order to determine the associated actions. These actions will consist of addressing a specific code repository and sending the code to specific execution environments in the network. The policy enforcement takes place in the Code Execution Controller Action Consumer. A high level example of this type of policy is as follows:

"*If* (invocation event $f(Ct_1)_1$ is received)
then (customized service must be downloaded as function of: $f(Ds_1)$ until $f(Ds_n)$ to IP addresses)"

Figure 5.7 represents how context information $f(\mathrm{Ct}_n)_n$ is represented as an event, which in turn triggers the execution of services as a function of $f(I_n)$. As result of the context variations defined by $f(\mathrm{Ct}_n)_m$, new invocations result in new context invocations, as defined by the function $f(I_{n+1})$, which in turn are used to define new services. In these functions, *n* is the index number to identify the type of context, while *m* is the number of samples or variations of same type of context information.

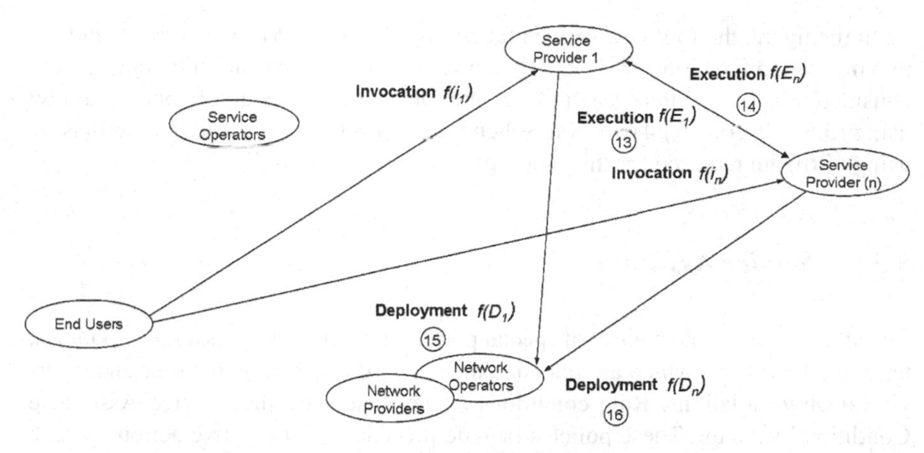

Fig. 5.8 Deployment of services as a result of service invocation

This figure presents four service functions $f(I_1), f(I_n), f(I_{n+1})$ and $f(I_m)$. In addition, two context functions $f(\mathrm{Ct}_n)_m$ and $f(\mathrm{Ct}_{n+1})_m$ are considered. As before, no context variations and only one type of information are assumed. Thus, (5.6) once again is used to determine that four ((1)(4) = 4) schema elements (e.g. semantic interactions) are required for this ontology representation.

5.3.5 Service Execution

Code execution policies, named *CExecution* policies, will govern how the service code is executed. This means that the decision about where to execute the service code is based on one or more factors (e.g. using performance data monitored from different network nodes, or based on one or more context parameters, such as location or user identity). The typical components with the capability to execute these activities are commonly named Service Assurance Action Consumers, which evaluate network conditions. Enforcement of these policies will be the responsibility of the components that are typically called Code Execution Controller Action Consumers. A high level example of this type of policies is as follows:
"*If* (invocation event $f(I_{n+1})$ is received or $f(I_{n+1})$ is TRUE)
then (customized service must be deployed as function of $f(D_n)$"

Figure 5.8 represents the deployment of service code. The context information is defined as part of the invocation function $f(I_n)$, and results in the triggering of services to be executed according to the function $f(E_n)$. In these functions, n is the index number identifying the type of context, and m is the number of samples or variations of the same type of context information from the source. As a result of context variations $f(\mathrm{Ct}_n)_m$, new deployment results in code executions controlled by the functions $f(E_1)$ through $f(E_n)$, and also results in deploying new services as $f(D_n)$.

In the figure, the four new service functions $f(E_1)$, $f(E_n)$, $f(D_1)$, and $f(D_n)$, and the two invocation functions $f(I_1)$ and $f(I_n)$. However, the four invocation functions were considered in the previous graph representation. Thus, in this graph, once again by using (5.6), observe (1)(4) = 4, four schema elements (e.g. semantic interactions by using (5.6)) are required for this ontology.

5.3.6 Service Assurance

This phase is under the control of special policies termed service assurance policies, termed *SAssurance*, which are intended to specify the system behaviour under service quality violations. Rule conditions are evaluated by the Service Assurance Condition Evaluator. These policies include preventive or proactive actions, which will be enforced by the component typically called the Service Assurance Action Consumer. Information consistency and completeness is guaranteed by a policy-driven system, which is assumed to reside in the service creation and customization framework. Examples of this type of policies are:

"*If* (customized service is running as a result of a deployment function $f(D_n)$) *then* (configure assurance parameters $f(A_{1+n})$ service & (configure assurance variables $f(D_{1+n})$)"

"*If* ($f(E_n)$ = 1)&(parameterA > X) *then* (Action defined by function $f(E_n)$)"
"*If* ($f(D_n)$ = 1)&(parameterB > Y) *then* (Action defined by function $f(D_n)$)"
"*If* ($f(A_n)$ = 1)&(parameterC < Z) *then* (ParameterA > X)&(Action defined by function $f(D_{1+n})$)"

Specifically, in this phase, the externally provided information can either match pre-defined schema elements to achieve certain management activities or, more importantly, the management systems can use these schema elements to extend and share the information to other management systems.

This extension requires machine-based reasoning to determine the semantics and relationships between the new data and the previously modelled data. Reasoning about such decisions using ontologies is relatively new; an overview of this complex task is contained in [Keeney06].

Figure 5.9 shows five service functions that make up part of the complete set of service management operations $f(A_1)$, $f(E_{1+n})$, $f(E_{1+m})$, $f(D_{1+n})$ and $f(D_{1+m})$. The context information functions are considered atomic units; thus, in this graph, again the (5.6) is used to determine the required five ((1)(5) = 5) schema elements (e.g. semantic interactions).

Note that if all schema elements from Sects. 5.3.2–5.3.6 are added together, thus it is obtained 21 total schema elements that are required to represent service management operations. The same value is obtained when solved the semantic-based logic function shown in (5.10).

Figure 5.9 represents the service assurance aspect of this approach. It shows associations of context information $f(\mathrm{Ct}_n)_m$ through $f(Ct_{1+n})_{1+m}$, which can be

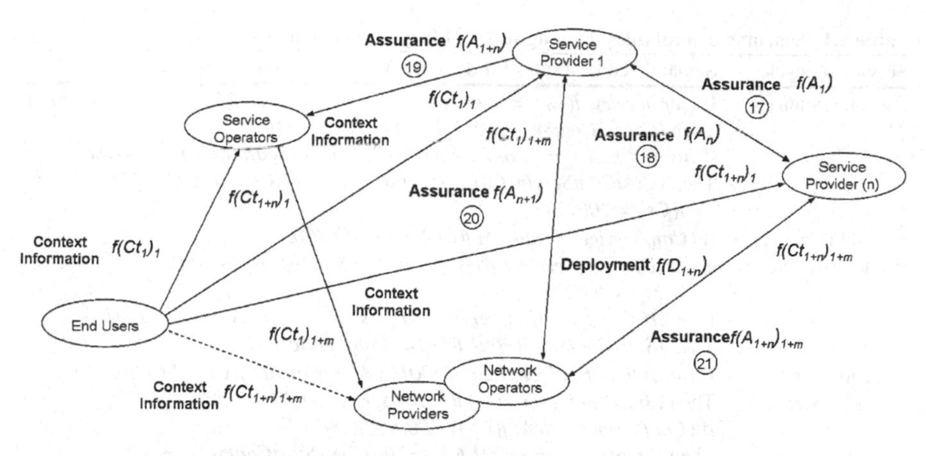

Fig. 5.9 Context information controlling service assurance

expressed using policy conditions and actions that use information coming from the external environment for executing and controlling the management operations for ensuring the service.

5.4 Ontology-Based Operations Using Semantic-Based Rules

This section presents research advances on semantic rules for combining service management operations and context information models for promoting information interoperability within pervasive applications. This section briefly describes a novel management technique using context information and ontology-based data models. Semantic rules associated with management operation functions are now presented. These semantic rules act as the foundation for building ontologies by expressing these functions as semantic interactions.

The control rules used to resume the service lifecycle functions follow logic-based rules that are expressed as policies, which defines a set of actions to be executed when certain conditions are met. This can be expressed using If-then-Else type rules.

Table 5.1 shows a set of function-based rules expressed as Condition-Action policies for managing the service lifecycle operations; these rules are expressed in OWL. This table shows how policies can be created using semantic functions (i.e. this table shows a function-based description of service management operations using policies).

When the values of the semantic functions match concepts from the ontology, the semantic functions can be used for controlling the service management operations. These can be used to orchestrate a set of policies, and as the context information need to be integrated in those management operations, the context information

Table 5.1 Semantic control rules defining service lifecycle managing functions

Service lifecycle	Semantic control rules for defining ontology-based functions
Customization	If (*serviceNew* $f(Cr_n)$ = "*Service001*") & (*LocatedIn* $f(Ct_n)_m$ = "*WebServer*") Then (*CreateConfService001* $f(Cr_n)$= "*Service001*") If (*userOf* $f(Ct_n)_m$ = "*ConfService001*") & (*LocatedIn* $f(Ct_n)_m$ = "*NetServer*") Then (*CustConfServ001* $f(Cs_n)$ = "*001*") & (*CustConfServ002* $f(Cs_n)$ = "*002*")
Distribution and deployment	If (*ConfServiceScheduleAt* $f(Ct_n)$ = "*00:00:0000*") Then (*DistConfServ001* $f(Ds_n)$ = "*001*") & (*DistConfServ002* $f(Ds_n)$ = "*002*") If (*userOf* $f(Ct_n)$= "*ConfService001*") & (*locatedIn* $f(Ct_n)$= "*Reg001*") Then (*StartConfService001* $f(D_n)$= "*ConfService001*")
Execution and maintenance	If (*userOf* $f(Ct_n)$= "*ConfService001*") & (*LocatedIn* $f(Ct_n)$= "*Cell001*") Then (*StartConfService002* $f(E_n)$= "*Service002*") If (*ConfServiceDateAt* $f(Ct_n)$= "*00:00:0000*") Then (*StopConfService001* $f(A_n)$= "*001*")&(*StartConfService002* $f(A_n)$= "*002*")

is then assigned to the values of the mentioned ontology-based functions. In this way, the context information models, represented by formal languages like OWL, can support service management and provide information interoperability leading to self-management in autonomic communications.

A novel characteristic when ontologies are introduced in service management operations and systems is the integration and harmony between context information awareness and policy-based management. Ontologies aim to solve one of the main problems in the management of services and networks, which is integrating context information in tasks for managing networks service operations. In particular, ontologies following function-based rules can be used to detect context changes, and hence change the functionality offered by the system in response to the new context.

5.5 Conclusions

In this chapter…

Pervasive service lifecycle operations have been described and represented. It has been demonstrated that it facilitates the formal representation of service management operations and the different phases of the service lifecycle have been explained. New organizational aspects of emerging Internet services and communications (ITC) demands, as well as network management operations and cloud computing functions in service modelling tasks, were being considered when this task was presented.

Management operations for pervasive services have been explained, emphasizing six important phases—creation, customization, management, operation, billing and customer support. The semantic control of service management operations is the focus of this chapter. Five types of service management operations (code distribution, code maintenance, invocation, code execution and assurance) were examined with respect to being able to support semantic variations of management and context information, so that application-specific behaviour can be orchestrated.

The use of functions following set theory and semantic rules for controlling the management operations of the service lifecycle has been exposed. The solution seems suitable for any particular technology, as it is following self-management principles inspired from autonomic communications. The comparative results between a management system using policies with and without semantic enrichment and ontology-based functions illustrate the differences of using this approach.

An alternative for formal representation about the service lifecycle operations has been described. Formal representations of service management operations for supporting the information interoperability in pervasive environments are considered as well. Finally, it is demonstrated that service management operations are described and formally represented with the combination of ontologies and information models following the principles explained in this chapter.

Chapter 6
Ontologies for Cloud Service and Network Management Operations

6.1 Introduction

This chapter presents evaluation result about the general concept of using ontology engineering to represent network management operations and cloud services. The results are introduced as experiences and descriptions, scenario descriptions on service and network management, autonomic architectures, service application frameworks and cloud computing applications. The scenarios help as reference and examples with the aim of showing how ontologies contributes for supporting pervasive services and network management systems in cloud environments following autonomic principles.

The aim of this chapter is to depict autonomic applications for providing management and support of services that require certain levels of "smart" understanding (which in turn enables ontology engineering processes like semantic reasoning, for example or other ontology engineering techniques) when context information is integrated into the service management application. This chapter also describes how ontology engineering is being used for the interaction of applications for management operations of pervasive services in cloud environments.

The organization of this chapter is as follows. Section 6.2 summarizes the service management benefits using ontology engineering, if well it is not an exhaustive analysis but it describes the most general benefits when ontology engineering is used in service and network management, autonomic and cloud environments.

Section 6.3 evaluates the general idea of using ontologies and the mechanisms or tools to facilitate integration of models and interoperability. This section concentrates more in the formal representation about the integration of information in pervasive services management rather than software tools or technology factors for facilitating the information interoperability as in previous chapters in this book.

Section 6.4 surveys research challenges and features that define information model interactions that are required to achieve the level of information interoperability that is required for next generation networks (NGNs) supporting services that use autonomic communications principles and mechanisms.

J.M. Serrano Orozco, *Applied Ontology Engineering in Cloud Services, Networks and Management Systems*, DOI 10.1007/978-1-4614-2236-5_6,

Section 6.5 introduces a functional framework in the form of components approach named PRIMO (policy relations and interaction for management using ontologies). This functional framework depicts how using the key concepts explained in this book (i.e. context-awareness, policy-based management, ontology engineering and cloud principles) pervasive applications can be managed accordingly with service-user demands. The objective of this architecture is to represent a realistic autonomic communications scenario, where the ontologies acquire application-specific values in the process of integrating context information into the management operations of pervasive services.

Section 6.6 introduces representative scenarios for the Internet and generic application management systems. These scenarios use ontologies to create interoperability across different management domains using semantic reasoning mechanisms that leverage policy-based management mechanisms and achieve autonomic behaviours. The scenarios inspired by seamless mobility are presented for providing insight on future directions and applications in both the cross-layer interoperability of NGN systems and the integration of context for service management operations and cloud computing environments.

Section 6.7 introduces scenarios highlighting trends towards the convergence of networks and services where multiple and diverse networks and technologies to interoperate are involved; the emerging business-driven architectures and management systems in the cloud are also discussed in this section. This scenario concentrates in virtual infrastructures to help for understanding about the service lifecycle control and thus promotes ontology engineering and autonomic technologies as an emerging way towards managing next generation cloud management services.

Section 6.8 presents the conclusions of this chapter.

6.2 Service Management Benefits Using Ontology Engineering

The Internet services and systems, from a design conception, does not pursue free information exchange between networking data and service levels; mainly because the model which was created did not pursue that aim. Currently, the Internet is by many meanings a complete different platform beyond the only objective of packet switching network between universities and government office, [Fritz90], today the Internet has a face with clear business objectives, thus it is wise to re-consider the role of the Internet in the services and systems and re-design utilities targeting this requirement. This fact promotes a race between academic and industry communities, first to investigate for designing what it could be a solution enabling this feature and second to bring to the market products enabling this feature and enhancing the current solutions with emerging novelty products. This new design feature has contributed in the transformation from an agnostic to a more service and network aware of Internet services, and as an inherent consequence, communication systems are following this evolution line too.

In this accelerated necessity for designing the Future Internet services and systems, many active academic and information technologies and communication's (ITC) industry communities have participated with approaches to enable information interoperability proposing a new design conception [NEWARCH], [Blumenthal01], [Feldman07]. Unfortunately, realistically this evolution in designing has not promoted a coordinated course in terms of implemented solutions, due to many complex issues involving deployment of diverse software solutions and the fast evolution, in parallel, of multiple heterogeneous infrastructures.

It is expected, looking at the future of the Internet in a service-oriented manner, where a more consolidated service-oriented role is deployed and where link information co-exist, could satisfy part of the many complex challenges. This view facilitates integrated tasks of services and networks following common guidelines to provide solutions in the form of implemented and interoperable mechanisms.

Challenges in the future communications systems, mainly Internet-based systems, demand, in terms of end-user requirements, personalized provisioning, service-oriented performance and service-awareness networking, supporting the requirements of information interoperability and data model integration [NGN]. Following, this short list of essential benefits is introduced:

1. Information can be shared easily between different skate holders in organizations and systems.
2. Users can be offered with variety and wide numbers of composed services, which they can personalize to meet their evolving needs.
3. Communities of users can tailor particular services to help create, improve and sustain their social interactions.
4. The services offered can be service-driven rather than technology-driven, however, designed to maximize the usage of capabilities of underlying technologies.
5. Services can be designed to satisfy user requirements per se, and thus be readily adapted to their changing operational information/data.
6. Service providers can configure their infrastructure to operate effectively in the face of changing service usage patterns and technology deployment.
7. Integrated solutions can be planned, designed and implemented with the common goal of managing different aspects of the information with no concerns about where the information flows will be processing.
8. Infrastructure can be optimized, on a group-basis, to meet specific low-level objectives, often resulting in optimal operation considering mainly important business and service user objectives.
9. Infrastructure and devices can be designed following specific and interoperable shared-based and data-linked standards.

6.2.1 Advances on Information Interoperability in the Future Internet

Convergence between technologies for communications networks (communications) and Internet services (Software) has been a clear trend in the information and communications technology (ICT) area in the past years, as it has been discussed already in Chap. 1 of this book. This trend involves, mainly, services management issues between non-interoperable aspects in software systems and communications infrastructure. In this panorama, perhaps a more consistently orientation for understanding how ontology engineering can support the new service, and communication challenges result more interesting than new definitions or procedures aiming to guarantee such benefits.

Research initiatives addressing software-oriented architectures (SOA) base their implementation in overlay networks that can meet various requirements whilst keeping a very simplistic, unmanaged underlying Internet platform (mainly IP-based). For example, a clear example of this SOA design orientation is GENI NSF-funded initiative to rebuild the Internet [NSFFI]. Others initiatives argue about the requirements for a fundamental redesign of the core Internet protocols themselves [CLEANSLATE], [NGN].

As the move towards convergence of communications networks and a more extended service-oriented architecture design gains momentum worldwide; facilitated mainly by pervasive deployment of Internet protocol suites (VoIP is a clear example of this), the academic research community is increasingly focusing on how to evolve networking technologies to enable the "Future Internet" and its implicit advantages. In this sense, addressing evolution of networking technologies in isolation is not enough; instead, it is necessary to take a holistic view of the evolution of communications services and the requirements they will place on the heterogeneous communications infrastructure over which they are delivered [IFIF], [SFIFAME].

By addressing information interoperability challenge issues, Internet systems must be able to exchange information and customize their services. So Future Internet can reflect changing individual and societal preferences in network and services and can be effectively managed to ensure delivery of critical services in a services-aware design view with general infrastructure challenges. A current activity in many research and development communities is the composition of data models for enabling information management control. It focuses in the semantic enrichment task of the management information described in both enterprise and networking data models with ontological data to provide an extensible, reusable, common and manageable data link plane, also referenced as inference plane [Serrano09].

6.2.2 Benefits on Interoperability and Linked-Data by Using Ontology Engineering

Taking a broad view of state-of-the-art, current development about ontology engineering and particularly data link interactions in the converging communications and software IT technologies, including trends in cloud computing, many of the problems present in current Internet systems and information management systems are generated by interoperability issues.

Ontology engineering provides tools to integrate user data with the management service operations, and offers a more complete understanding of user's contents based on data links and even relating their social relationships. Hence, it is positive a more inclusive governance of the management of resources, devices, networks, systems and services can be used for supporting linked-data of integrated management information within different management systems. This complete understanding of contents use ontologies as the mechanism to generate a formal description, which represents the collection and formal representation for network management data models and endow such models with the necessary semantic richness and formalisms to represent different types of information needed to be integrated in network management operations. Using a formal methodology the user's contents represent values used in various service management operations, thus the knowledge-based approach over the inference plane [Strassner07b] aims to be a solution that uses ontologies to support interoperability and extensibility required in the systems handling end-user contents for pervasive applications [Serrano09].

6.3 Integration of Models by Using Ontology Engineering

Trends in the area of ITC require multiple and diverse networks and technologies to interoperate and provide the necessary support and efficient management of pervasive services. In this sense, autonomic system principles and emerging cloud technologies and their management tools are an important way to facilitate and enable solutions capable for managing the complexity of NGN services. This section discusses ontology engineering enabling information interoperability, and outlines an approach for lifecycle management of services and applications that require specific levels of quality of service (QoS) as particular example to show the advantages of ontology engineering to achieve this objective. Ontology engineering is used to enable the interoperability across different management domains using semantic reasoning and leveraging policy-based management techniques to achieve autonomic behaviour.

In recent years, the business and technical aspects, and hence the complexity, of Internet and communications services and their support systems have increased enormously, requiring new technologies, paradigms and functionality to be introduced. The drive for more functionality has dramatically increased the complexity of the systems so much, so that it is now almost impossible for a human to manage all of the different operational scenarios that are possible in today's complex communication systems.

The common stovepipe present in the design of OSSs (operational support systems) and BSSs (business support systems) exemplify the necessity to incorporate the best implementation of a given feature; unfortunately, this focuses on the needs of individual applications as opposed to overall system needs, and thus inhibits the sharing and reuse of common data. This is a specific example of the general inability of current management systems to address the increase in operational and business complexity by being unable to share and reuse what should be common network and management data.

Autonomic solutions are designed to provide self-management capabilities to systems [Strassner06a], and were designed to manage the growing business and technical complexity in system management. Closely to autonomic solutions the policy-based paradigm is related. The policy-based management paradigm has been used by several approaches to tackle the problem of fast and customizable delivery for different types of networking solutions [Aidarous97]. QoS definition, emphasizing information interoperability, has been examined in OPES [OPES] and E-Services [Piccinelli01]. However, new tools are required that enables business perspectives to be modelled and used to determine network services and resources. Policy-based systems applied in service management are used for governing the behaviour of a system separate from the information functionality that the system provides. In other words, policy management solutions provide the ability to govern operations and adapt the services and resources that the network offers in response to changing user demands, business requirements and environmental conditions. This is one of the inherent benefits of autonomic mechanisms, since autonomic computing and networking had these goals in mind from the beginning [Kephart03], [Strassner06a]. Today, trends make use of middleware platforms to achieve context information interoperability necessary when information is being shared, and thus offer a solution for complex system management integrating such information.

Here after, this section is primarily concerned with using information and policies to manage communications services in autonomic environments and/or NGNs, as it is a main goal specified in many up-to-date initiatives, including [OPES], [IFIP-WGS] and [IFIP-MNDSWG]. Up to date policy-based management for networks and services has focused on developing languages and architectures for managing and deploying policies in distributed environments. However, many open issues exist in this domain, such as policy conflict detection and resolution [Davy08a], [Davy08b], along with policy refinement and policy reasoning [Davy07b].

The current state-of-the-art in policy management dictates targeted solutions for different functions, but the most of them concentrate on the management of networking applications [Schönwälder99]. For example, if security and network configuration are both desired to be governed using policies, the usual solution is to deploy separate policy servers [Strassner06a]. A service-oriented solution complicates system governance, since the interaction between policies from different contexts and domains (e.g. service and network) become more complex due to the lack of a common network management *lingua franca*.

In this framework, domain interactions using ontologies can be depicted as group interactions and represents the relationships between the classes from the different ontology-based model representations, as shown in Fig. 6.1. For example, the context information model, the policy information model and the service management operations, all them have different relationships that represent these class interactions. The ontology-based models contain class correspondences, and the interactions are used to construct ontology interactions areas as shown.

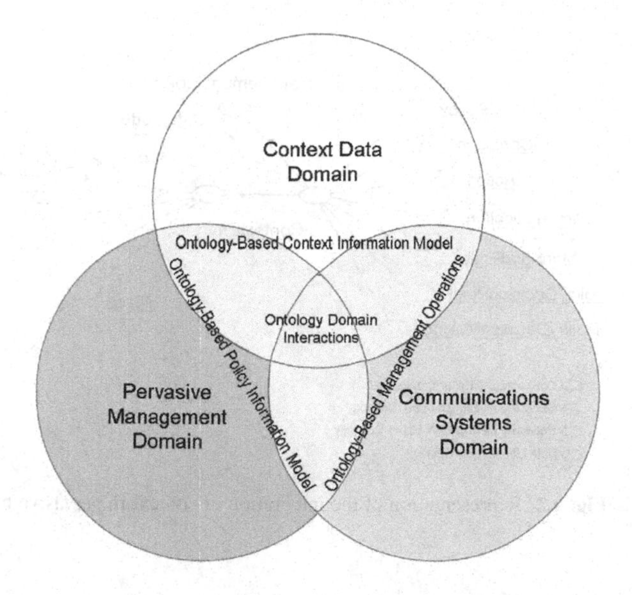

Fig. 6.1 Ontology domain interactions—upper ontology representation

The objective of defining the class interactions is to identify classes whose content must be integrated with all or parts of the information necessary to be shared. The task of identifying these interactions is done with a visual approach using the ontology class diagrams based on the individual information model class definitions that have been previously presented and explained in the previous chapter of this book.

The class diagrams contain the class relationships described in a generic form (e.g. as abstract classes interacting for deploying a service), but each relationship and interaction between classes is appropriately refined when the ontology is constructed and subsequently edited. Thus, the class interactions map becomes a tool for representing the semantic interactions which make up the ontology. In other words, the ontology becomes a class map to identify inter-linked concepts

It is important to highlight the difference between relationship and interaction:

Interaction—The graphical representation for two or more elements in class diagrams that interact and hence define one or more specific behaviours. This is normally shown using symbols that represents the type of interaction, so that different interactions having different meanings can be easily identified.

Relationship—The semantic description of the interaction between two or more elements, usually consisting of a description that follows simple syntax rules and uses specific keywords (e.g. "*isUsedFor*" or "*asPartOf*").

Figure 6.2 shows the domain ontology representation (upper-representation). The image represents the integration of the context information classes related to the management operation class through the Event class. The Event class interacts with other classes from different domains in order to represent context information. Note that only the ContextEntity class from the context information model domain

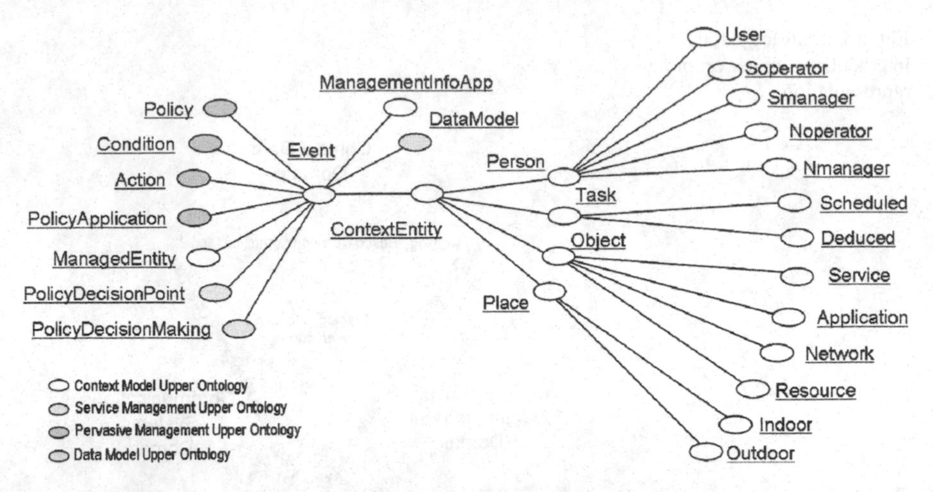

Fig. 6.2 Representation of the integration of context in pervasive management

and the Event class from the service management domain are shown. This simplifies the identification of interactions between these information models. These entity concepts, and their mutual relationships, are represented as lines in Fig. 6.2. For instance, a ContextEntity forms part of an Event class, and then the Event defines part of the requirement evaluating a Condition as true or false. Another example shows that one or more Policies govern the functionality of a Managed Entity by taking into account context information contained in Events. This functionality enables context to change the operation requested from a pervasive service or application, and is represented as the interaction between Event and ContextEntity.

The ontology seen in this class representation, integrates concepts from the IETF policy standards as well as the current DEN-ng model. Thus, in the following subsections, important classes that were originally defined in the IETF, DEN-ng, models will be identified as such. The ontology defines a set of interactions between the Context Data, Pervasive Management and Communications Systems Domains in order to define the relationships that represent interactions between the classes from the information models for these three different domains. These interactions are an important part of the formal lexicon, as shown in Fig. 6.3 and will be described.

The formal language used to build the ontology is the web ontology language (OWL), which has been extended in order to apply to pervasive computing environments [Chen03c]; these additional formal definitions act as complementary parts of the lexicon. The formal descriptions about the terminology related with management domain are included to build and enrich the proposal for integrating network and other management data with context information to more completely define the appropriate management operations using formal descriptions. The ontology integrates concepts from policy-based management systems [Sloman94b], [Strassner04] to define a context-aware system that is managed by policies, which is an innovative aspect when integration work is being performed.

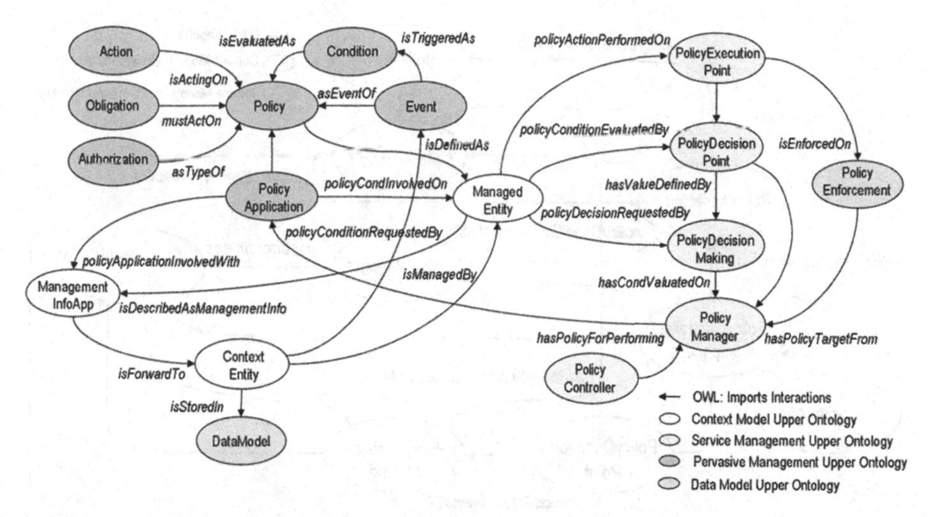

Fig. 6.3 Domain interactions—class ontology representation

From Fig. 6.3, ManagedEntity is the superclass for Products, Resources and Services, and it defines the business as well as the technical characteristics for each. A PolicyApplication is a type of Application. It evaluates a set of PolicyConditions and, based on the result of the evaluation, chooses one or more PolicyActions to execute. A PolicyController is equivalent of a "Policy Server" from other models; the name was changed to emphasize the fact that PolicyApplications are not limited in implementation to the traditional client–server model. The ontology class interaction *hasPolicyForPerforming* signals when the set of PolicyRules, selected by the current ContextData, are ready to be executed. When this interaction is instantiate, the PolicyController can pass the set of policies to the PolicyManager, which is a class in the ontology. The ontology is based on superclass interactions between the classes from the domains involved in the integration of context information for pervasive service management operations. This simple example about ontology interactions shows representation and the identification for domain interactions (see Chap. 5 for more details about various pervasive service management operations).

6.4 Functional Blocks Architecture Supporting Information Interoperability

Before introducing the architecture and describing the basic components, it is important to highlight the concept of policy interactions. In network communication systems, policy interactions are the concept of linking the effects of one policy to trigger other policies, independent of the type or abstraction level of the policy.

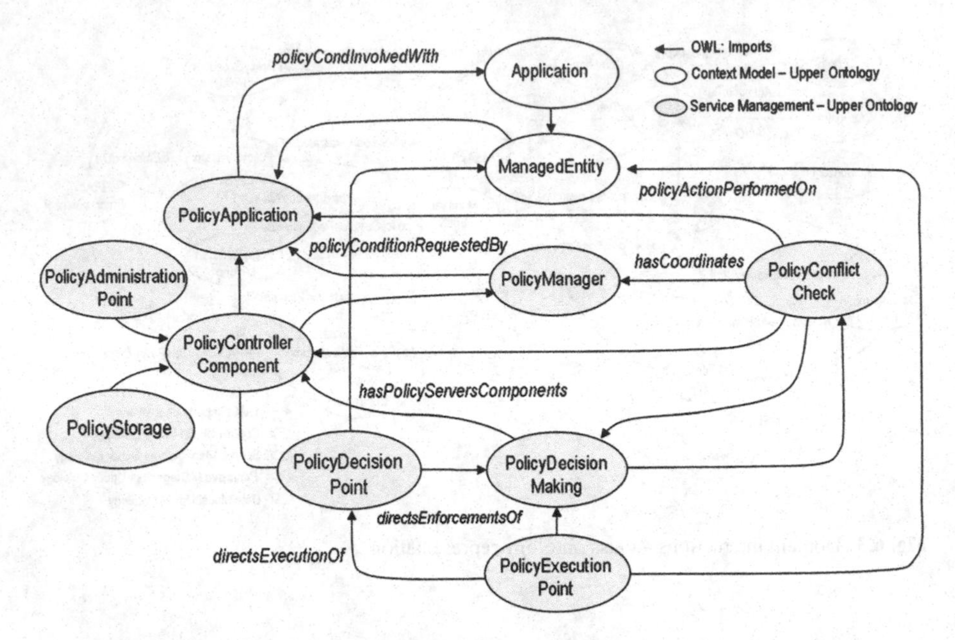

Fig. 6.4 Onto-CONTEXT architecture class representation for service management

For example, there is a fundamental difference between an obligation policy and an authorization policy: the former defines what actions a subject is permitted (or not permitted) to perform on a target, and hence is target-based enforcement; the latter defines the actions a subject must do to a target, and hence is subject-based interpretation and enforcement [Strassner08] and [Damianou01], respectively.

As related to this chapter's section,the architecture supporting information interoperability is introduced in Fig. 6.4. The functional blocks and components are described in detail later in this chapter. This functional architecture uses policies and ontologies as the formal mechanism for representing and exchanging information. In this functional architecture, policies are defined for service lifecycle management from the perspective of the service provider interacting with policies pre-defined (e.g. policies for traffic engineering from the perspective of the network provider). This example is different from policy refinement, where policies at one level are derived from policies at a higher level or vice versa (e.g. policies controlling services are used to derive policies that control the network that provides those services).

The proposed functional architecture is focused on solutions that enable multiple policy-based systems that use their own, different, policy models to interoperate with each other. This is done by using ontologies and semantic integration, and it is triggered by the events signalled between the involved domains and the policies at all appropriate abstraction levels.

Policies can exist at multiple abstraction levels, such as administrative domains—this is codified as the Policy Continuum [Strassner04] [Davy07a]. Policy interaction is the explicit triggering of policies across management boundaries and abstraction levels in order to achieve a specific set of goals and/or behaviours. Note that since policies are fundamentally related to each other and they are essential part in autonomic solutions; as expressed in the Policy Continuum, policies from each interacting management domain can affect each other.

The combination of policies and context information in a formal way has been proposed for supporting and managing services [Serrano06c]. In this chapter's section, a step further is taken to briefly explain and discuss an architecture that is intended to support high-level policy service interactions for controlling the network traffic from many diverse stakeholders, including content providers, service providers and network operators. Figure 6.4 shows the service management components representation and the associated management operations involved in the service lifecycle process. Those operations are typically represented as policy management concepts when UML class diagrams are being used. However, in this time, a novel vision using class interaction maps is shown to represent these interactions, and is explained shortly as follows.

The main operations that a policy-based system can execute on a ManagedEntity are shown in the class interactions map, the novel use of this interaction map enables the visualization of the semantic relationships necessary when different classes are being related to each other where the PolicyManager works to satisfy the *policyConditionsRequestedBy* relationship between the PolicyManager and the PolicyApplication.

The PolicyApplication is related to the ManagedEntity in several ways, including directing which management and/or context information is required at any given time. Since a ManagedEntity is directly related to different Applications, those Applications can act as brokers or wrappers for service provisioning. In addition, some information from the Application can be used by policies. The PolicyManager governs and coordinates monitoring as well as management decisions (and their enforcement) using various appropriate entities, such as PolicyExecutionPoints and PolicyDecisionPoints (PDPs). The PolicyApplication controls all policies in the system related with the Application to share and reuse information contained in the policies.

6.4.1 Onto-CONTEXT Framework Components Description

A PolicyManager is a fundamental class, and represents both a set of core functionality for implementing policies as well as a unit of distribution in a distributed implementation. The PolicyManager works in coordination with a set of specialized policy components, including the PolicyExecutionPoint, PDP and PolicyDecisionMaking classes. These specialized policy components enable application-specific functionality of a PolicyManager to be built based on the use of a set of particular policy-based management operations, many of which can be defined as reusable components.

A PDP is similar to traditional PDPs (such as that defined in the IETF), except that it is specifically designed to answer requests from policy-aware and policy-enabled network elements (as represented by the ManagedEntity class interaction *policyConditionEvaluatedBy*) using formal ontology-based terms. This enables a PDP to serve as an interface between the network and higher level business processes. The difference between a policy-aware and a policy-enabled entity is a longer discussion that is beyond the scope of this book. For now, it is enough to say that the semantic expressiveness reached when using the combination of policy-aware entities and ontology-based mechanisms for representing and integrating context makes this proposal different from traditional policy-enabled approaches. The PolicyExecutionPoint is used to execute a prescribed set of policy actions on a given ManagedEntity as the class interaction *policyActionsPerformedOn*. A PolicyEnforcementPoint is a class that performs the action on a given ManagedEntity using the class interaction *isEnforcedOn*.

The PolicyDecisionMaking entity received requests from a ManagedEntity and evaluates PolicyConditions (via the class interaction *hasCondValuatedOn*) based in part from the results of the PDP (as a result of the class interaction *hasValueDefinedBy*). The class interaction *policyConditionRequestedBy* tells the PolicyApplication to evaluate one or more specific PolicyConditions; this may depend on retrieving management information using the class interaction isDescribedAsManagementInfo (which involves the ManagementInfoApp and the ManagedEntity). The class interaction *isForwardTo* is established when it is necessary to operate on the ContextEntity to store the context information in the DataModel using the class interaction *isStoredIn*. In order to integrate context information properly, the *isDefinedAs* class interaction (between ContextEntity and Event) enables the Event to trigger the evaluation of a PolicyCondition, as it is part of the Condition values of the policy-aware information model. This is defined as the two class interactions *isTriggeredAs* and *isEvaluatedAs* (with the Condition and Policy components, respectively).

The PDP obtains the status of the ManagedEntity after the execution of the PolicyAction(s) using the *policyApplicationInvovedWith* and the *isDescribedAsManagementInfo* class interactions. The class interaction *hasPolicyTargetFrom* is used to check if the PolicyActions have correctly executed.

Figure 6.5 shows the full class map interactions as multiple classes from the three diverse domains (context information model, pervasive management models and communications networks). In the ontology map, a "Policy" can be triggered by "Events" that relate important occurrences of changes in the managed environment, such as context-related conditions. Note that the upper ontology can contain and relate concepts to create new concepts that can be used by other applications in a lower ontology level or specific domains. For instance, "Router" is a type of "Resource"; this concept is defined as an "Object" that can be part of the context (which is represented by "ContextEntity"). "Router" has relationships to other objects, such as the "IP" class, which is a type of "Network."

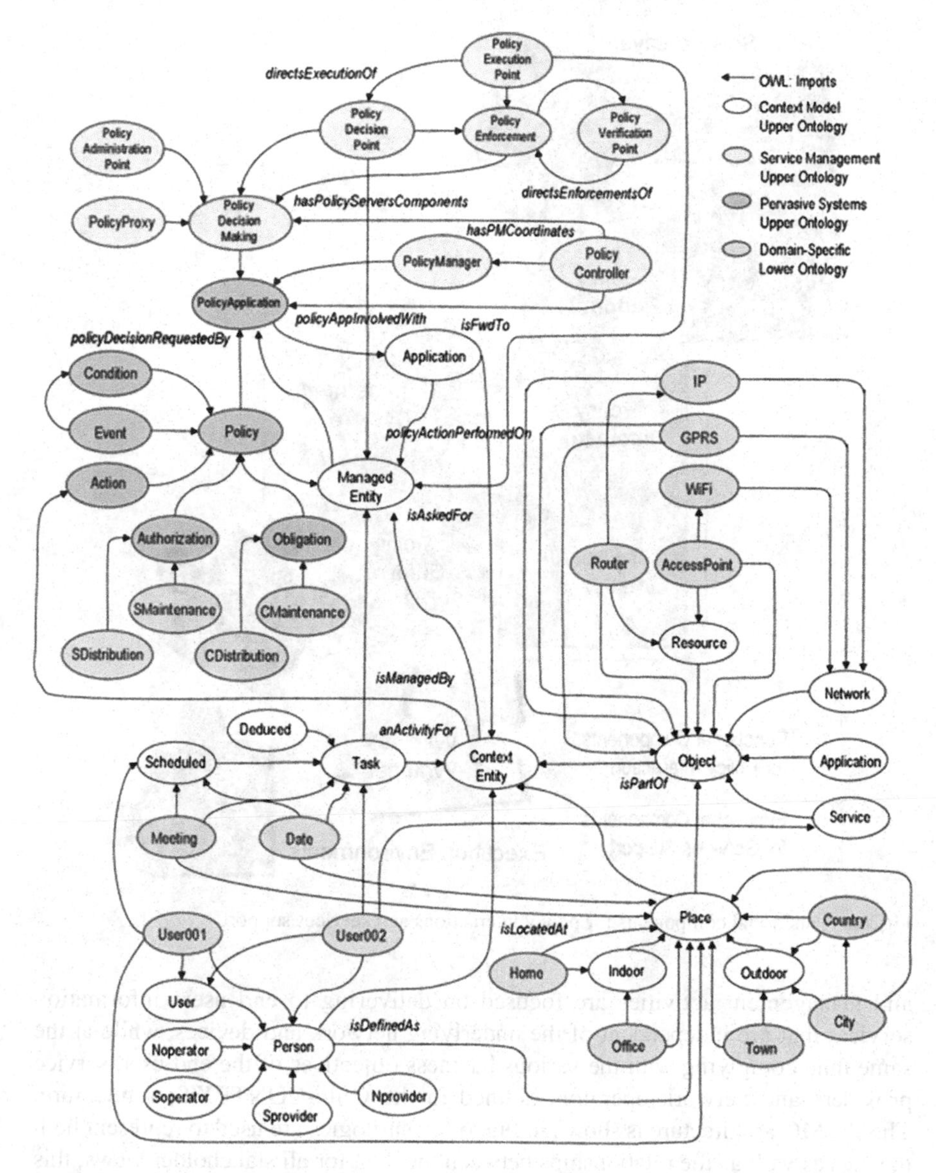

Fig. 6.5 Upper ontology extended representation

6.5 PRIMO Architecture—Information Model Interactions Middleware

An example of an architecture that uses the ontology engineering paradigm and offers information interoperability is PRIMO architecture. PRIMO is a functional set of components providing a conceptual platform where ontology-based operations

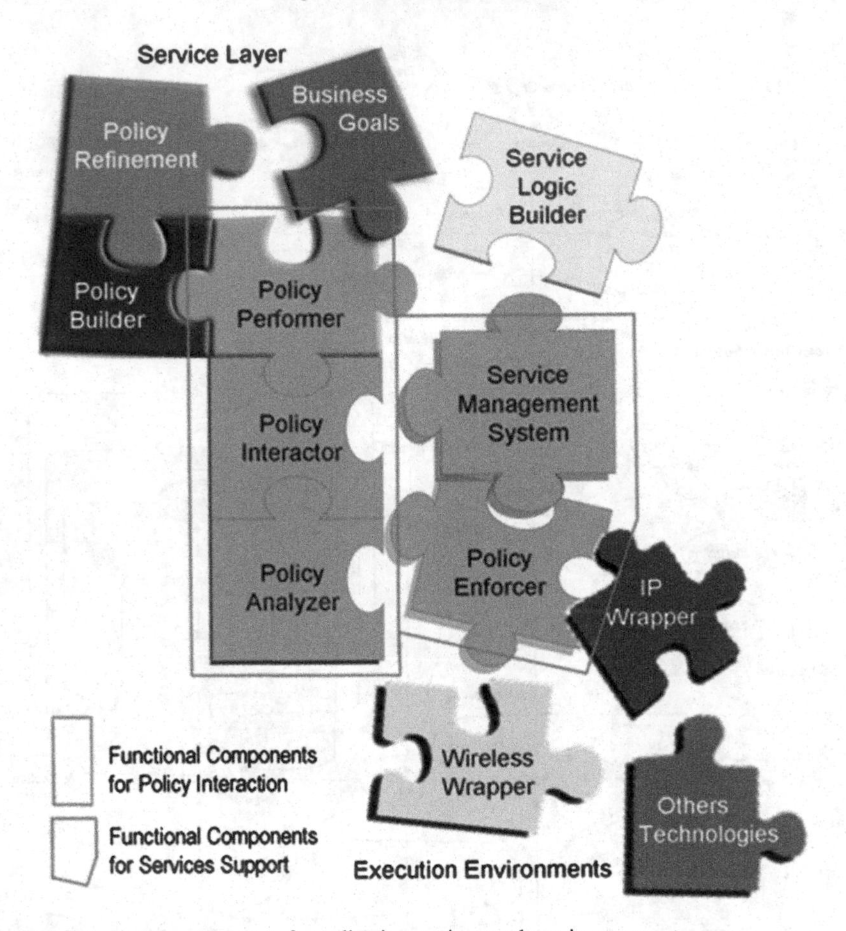

Fig. 6.6 Functional components for policy interactions and services support

and management activities are focused on delivering to end users information services that are independent of the underlying network and devices, while at the same time complying with the various business objectives of the end user, service providers and network operators defined in the Onto-CONTEXT architecture. The PRIMO architecture is shown in Fig. 6.6. Ontologies are used to represent both the data as well as the relationships between the data for all stakeholder views; this enables the interaction of various management views, especially those of the service provider and the network operator, to be modelled and, more importantly, reasoned about. This approach is outlined for policy transformation and interaction between different management systems showing how context information, represented via ontologies, is a facilitator for this approach. For more details visit [Davy07b] and [Serrano06a].

In PRIMO architecture the key elements involved in the policy interactions and described in this section are the Policy Performer, Policy Analyzer and the Ontology-Based Policy Interactor. These three components provide the framework for supporting policy interactions with other Policy-Based Service Management Systems. The core idea is to provide an intermediary system to act as a mediator between different policy-based systems and ontology-based solutions. If the architecture used is able to interact with ontology-based policy models, the exchange of information and the policy-based management activity can be done transparently as result of an information interoperability process using PRIMO.

PRIMO acts as an example of using ontology-based policy model for managing services in cross-layered environments because PRIMO uses ontologies in its definition and representation of various concepts. PRIMO uses the DEN-ng information model to provide a foundation for modelling appropriate entities (not just networking concepts, but business entities, end users, and others as well), and serves as a robust foundation from which ontology-based information models can be refined to support different application-specific needs. The PRIMO model extends the functionality of previous solutions for managing context-aware services, and acts as an extensible architecture that shows how ontologies can be used as a mechanism for facilitating information interoperability in autonomic solutions.

The PRIMO approach manages the full lifecycle of a service by using ontology-based policies, which represent the integration of policies with the desired semantics to be enforced. This integration is critical, since policies constructed from an information model lack the ability to reason (which of course ontologies provide), and enables the framework to take into account any variations in context information that are experienced (e.g. those contained in traffic engineering policies), and relate those variations to changes in the services operation and performance (e.g. QoS). When policy interactions, ontology-based data representations and policy-based management are used together, then a semantic platform that supports the engineering of autonomic systems can be provided.

The PRIMO proposal provides a platform for the validation of ontology models as can be seen, it is done for those components handling the ontology. The main objective of PRIMO is to deliver to end-users services that are context-aware and provide designated QoS services independent of any particular device or architecture, while complying with the underlying business objectives of the end users, service providers and network operators. The use of ontologies to represent both network and management data as well as relationships between those data for all stakeholders enable the interaction of various management views, especially those of the service provider and the network operator. The PRIMO architecture also defines an approach for policy transformation and interaction between different policy-based management systems, and shows how context information, represented via ontologies, is a critical facilitator for this activity.

The PRIMO architecture is based on the use of formal ontology languages, such as OWL [OWL]. The Policy Continuum, in conjunction with a set of associated ontologies, enables the translation from high-level goals (as codified by service policies) to low-level network view policies, and vice versa. This translation process is not a refinement process itself—rather, policies are transformed in content and representation. The context information is used only to relate policies at different abstractions to each other, and then use particular events to trigger and control each

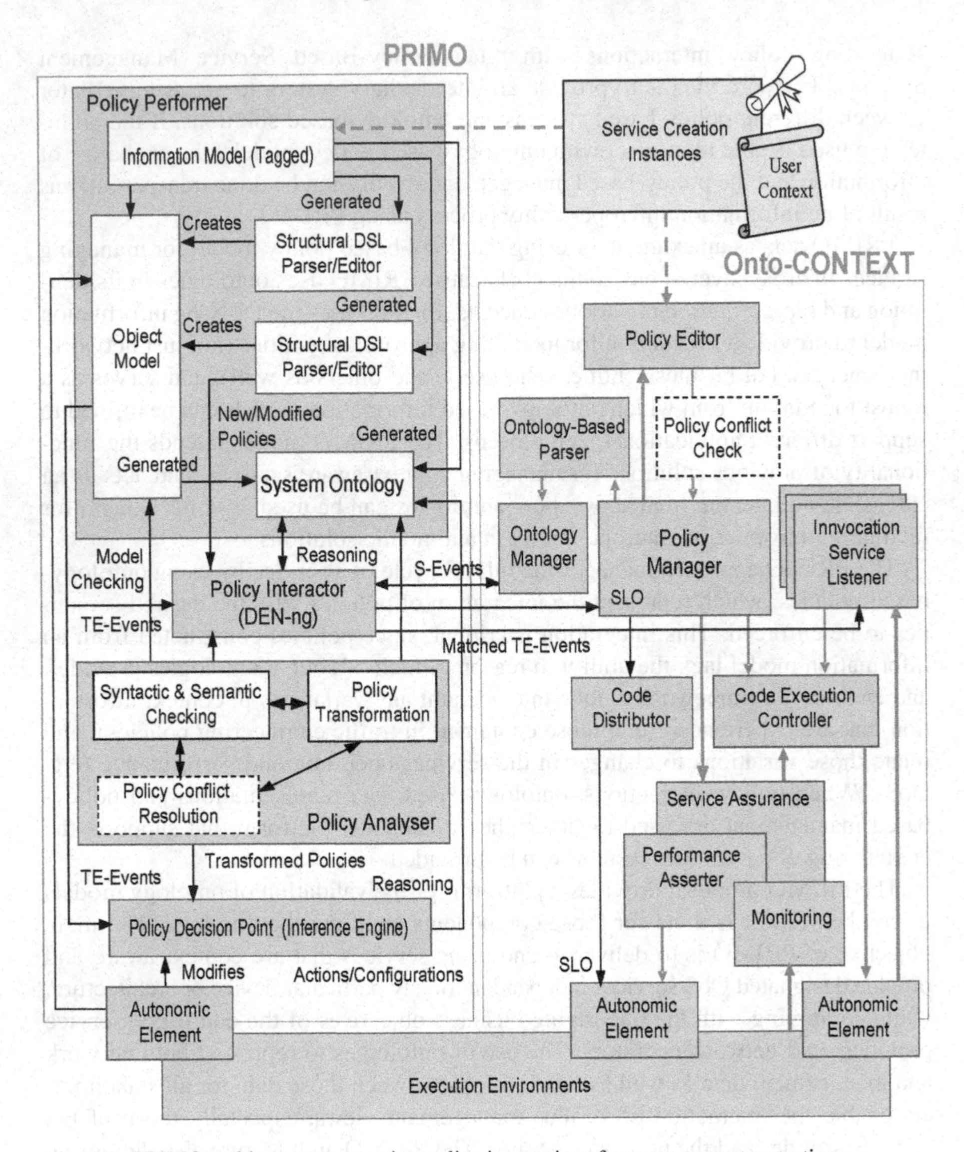

Fig. 6.7 PRIMO architecture supporting policy interactions for management operations

set of related policies. Thus, the component will take as input a service view policy, and select a set of network view policies that can meet its goals.

The selection process involves analyzing the triggering mechanisms of low-level policies as expressed in an appropriate set of ontologies via automated semantic integration, and matches them to the service level policy goals. An outline of this process is depicted in Fig. 6.7. The key enabling concept is the process of *semantic integration* that can create associations and relationships among entities specified in different ontologies and/or models.

The purpose of the PRIMO architecture is to manage traffic engineering conditions from a high level of abstraction. In the PRIMO architecture, the "Ontological Comparison" describes concepts and relationships between objects defined in ontology-based information models, which is based on the Onto-CONTEXT architecture, and the objects from the PRIMO components defined in the DEN-ng information model.

The use of PRIMO enables the Autonomic Manager to reason about received data and subsequently infers meaning. For example, PRIMO can determine which customers are affected by a particular SNMP alarm, even though customer information is not contained in the SNMP alarm, by looking at relationships in both the information model and the relevant ontologies; this enables PRIMO to infer which customers are affected by the SNMP alarm.

The autonomic system feature of PRIMO allows to define whether the actual state of the managed entity is equal to its desired state. If it is, then the system continues to monitor the managed entity. If it is not, then the system defines the set of configuration changes that should be sent to the set of managed entities that are not in the correct state, in order to govern their state transitions. Note that in general, one or more managed entities must be reconfigured in order to correct this problem. PRIMO can be viewed as intelligent middleware that translates vendor-specific data into a vendor-neutral form to enable the Autonomic Manager to operate on those data; then, this vendor-neutral form is translated into vendor-specific commands. This approach is, therefore, able to accommodate legacy as well as future devices.

The Autonomic Manager provides the novel ability to change the different control functions (such as what data to gather and how to gather it) to best suit the needs of the changing environment. An important, but future, area of research is to add learning and reasoning algorithms to the system. The former enables the system to grow and refine its knowledge base, while the latter provides the ability to generate hypotheses to more efficiently determine root cause. Finally, the Policy Manager is both an interface to the outside world (GUI and/or scripted) as well as a mechanism to translate those requests to the autonomic system. The framework is composed by components described briefly hereafter.

6.5.1 PRIMO Framework Components Description

Research into automatic semantic integration of ontologies is primarily concerned with the dynamic selection and invocation of web services based on a (usually) formal ontological description of what the web service provides. The key research challenge is to devise some methods of assurance that the service policies are actually satisfied by low-level policies when the semantic integration is automated across domain boundaries. A review of most popular approaches is discussed in [Damianou02] and, in the framework of autonomic solutions, in [IBM-PBM].

Table 6.1 Traffic engineering policies

POLICY_1	SUBJECT QoSAdmin TARGET Routers_on_PATH_A ON IPv4PacketRecieved IF IP source in 10.10.10.0/24 AND IP of type FTP AND IP destination in 10.10.24.0 / 24 THEN < MARK IP.ToS with AF41 >
POLICY_2	SUBJECT QoSAdmin TARGET Routers_on_PATH_A ON IPv4PacketRecieved IF IP source in 10.10.10.0/24 AND IP of type FTP AND IP destination in 10.10.24.0 / 24 THEN < SHAPE IP to 512Kbps >
POLICY_3	SUBJECT QoSAdmin TARGET Routers_on_PATH_A ON IPv4PacketRecieved IF IP.ToS equals AF41 THEN < CBWFQ weight is 5 >

As a framework example, suppose a service policy exists which specifies that when a new user account is created, a signal "CreateCustomerFlow" is forwarded as an event to the reasoner of the autonomic system. Further, suppose that the reasoner has pre-established, via semantic integration, that this event is equivalent to the set of network events IPAddressDetected, BeginTrafficIdentification and BeginTrafficConditioning. Each of these events can trigger related, lower-level events (e.g. BeginTrafficIdentification will trigger the classification and subsequent marking of the traffic). These events are initialized and forwarded to the network PDP for processing, and the relevant policies are installed and executed across the network. For each policy completed, an acknowledgement is returned to the PDP. The network PDP can forward a set of signals to the service provider PDP, informing it that it has triggered a number of policies. A reasoner, which is located in the service provider network, can intercept these events and match them against an equivalent service view event that is subsequently instantiated and forwarded to its PDP, signalling that the goal has been met.

The management of the services is done using a policy-based approach that uses context information from low-level policies to trigger service management operations using appropriate events. These events then trigger the appropriate high-level policies, so that business goals can be enforced in the network.

Table 6.1 shows policy examples and their interaction with service management policies. It contains a set of traffic engineering policies (in a pseudo policy language) that identifies FTP flows originating from the IP sub-domain 10.10.10.0/24 and destined for the sub-domain of 10.10.24.0/24. Once the flow is identified and marked as a class, there is a conditioning policy that instructs the appropriate queuing behaviour to be associated.

6.5.2 *Policy Performer or Selector*

The ontology-based performer is the component in charge of producing the traffic engineering policies that will be managed by the external policy-based management system as a set of policies. These policies define a new object, called the service level object (SLO), that is associated with the particular service level agreement in force. In this way, the traffic engineering is managed using policies throughout the entire service lifecycle. This ensures that the added value of services that offer a particular level of QoS can still be guaranteed. The policy editor generates the policies for the network view as a set of events, conditions and actions, which are defined and dynamically assembled to provide the desired behaviour. It is then up to the ontology-based performer to build those policies that will provide the traffic engineering support for the appropriate network services.

6.5.3 *Ontology-Based Policy Interactor*

The Ontology-Based Policy Interactor is the component that is responsible for semantically matching and transforming events received from different views to events between the same or different service systems. The reasoning process is performed for policy analysis and policy interactions to ensure that the policy events are syntactically and semantically compatible. Using formal semantics, as defined with ontologies, automated reasoning is facilitated using the enhanced representation available from the use of ontologies. This enhanced representation also enables the exchange, integration and querying of data by annotating simple management data with formal semantics [Horrocks05]. Defining an ontological model of a complex system from scratch is complicated and time consuming. The aim, therefore, is to leverage existing management models in defining ontologies; this also facilitates the definition and processing of semantic similarity between different concepts in the ontologies. To achieve this feature, the baseline system ontology is defined from appropriate subsets of the DEN-ng information model for example.

6.5.4 *Policy Analyzer*

The Policy Analyzer takes the specified policies and transforms them into the PDP rule engine format as well as a format suitable for generation device configurations. Our baseline ontology enables this component to take advantage of automated reasoning in performing the policy transformation and policy conflict detection processes.

Policy transformation is defined to be the mapping necessary to relate one representation of policy to another representation of policy. Note that policy transformations are usually performed between policies at different levels of abstraction. Since the different representations of policies often require different concepts, semantic integration techniques are required in the policy transformation process. Automated reasoning is an integral part of semantic interoperability [Wong05]; it is used, for example, to infer relationships that have not been explicitly identified.

Policy Conflict Detection determines whether a conflict between policies can occur [Davy08a] [Davy08b]. Conceptually, this can happen when two or more policies are activated simultaneously that are enforcing contradictory management operations on the system. Often, policies do not lend themselves to the detailed analysis needed to determine if their specific actions are in conflict; hence, in this case, the system ontology enables the Policy Analyzer to reason about this information to detect conflicting management operations.

6.6 Application Scenarios

Next generation services need to be dynamically deployed. Such services can be triggered automatically using semantic functions that, for example execute when the user logs in, such as a video conference or the downloading of multimedia content that starts on demand. Other example in how a service can be triggered automatically using semantic functions is when a user arrives to a geographical area for providing better connectivity service conditions. Furthermore, such services need to be assigned the appropriate QoS, which depends on various user attributes (e.g. the type of applications that the user wants to use) as well as user-centric, context-sensitive attributes (e.g. this is a user that has just purchased a new service, or this is an established gold service user). These scenarios require technological support based on automatic mechanisms that allow them to react in a specific manner based on changing context, and using the same formal language to exchange, share and reuse the information.

This section demonstrates the semantic framework using realistic scenarios that each benefit from using Ontology-Based Information Models and ontology-based architectures, e.g. onto-CONTEXT Architecture. Each scenario reinforces how the combination of ontologies, policy-based management and context-awareness simplify the complexity in the management of NGN and services in the autonomic communications area. A brief outlook about the pervasive scenarios and its general objective, acting as introduction manner, is presented as follows, the full description and technical aspects are presented in subsequent sections of this chapter.

Scenario 1, video conference services with QoS guarantees for image quality, based on contextual information from users and networks, CACTUS (*C*ontext-*A*ware *C*onference *T*o yo*U* *S*ervice scenario model). CACTUS is an enhanced extended version derived from the Moving Campus Scenario, which has been demonstrated as part of the EU IST-CONTEXT Project. CACTUS takes advantage of the knowledge present in the context information from users and networks, which are integrated in ontology-based information models, to provide an advanced service to its consumers. CACTUS is fundamentally a different service than conventional services that do not use context information for their configuration or to trigger the deployment of new services. CACTUS is responsible for providing QoS guarantees for specific time periods in order to hold a video conference session among the members of a group.

Scenario 2, distribution of services on demand of multimedia contents, based on contextual information from devices and networks, MUSEUM-CAW (*MU*ltimedia *SE*rvices for *U*sers in *M*ovement—*C*ontext-*A*ware & *W*ireless scenario model). MUSEUM-CAW is extended by using semantics from the Crisis Helper Scenario which was demonstrated in the EU IST-CONTEXT Project. MUSEUM-CAW takes advantage of the knowledge contained in ontology-based information models, and compares that information to contextual information from the devices and networks used to provide an advanced multimedia service that is independent of specific access devices in wireless environments. Multimedia services are customized using the user and devices profiles for the deployment of multimedia services in an ad hoc fashion. MUSEUM-CAW is responsible for the distribution of multimedia contents, and guarantees the transmission among different hot spots in a wireless network.

Scenario 3, integrated services that are technology-independent and automatically deployed Sir-WALLACE (*S*ervice *i*ntegrated fo*r* *W*iFi *A*ccess, onto*L*ogy-*A*ssisted and *C*ontext without *E*rrors scenario model). Sir-WALLACE is an enhanced version by using semantics from the Services to all People scenario which has been demonstrated in the EU IST-CONTEXT Project. This project exhibits self-management capabilities based on the use of ontologies, context-awareness and policy-based management. This scenario takes advantage of the knowledge in the context information from users, devices, networks and the services themselves, which are all integrated in ontology-based information models. Sir-WALLACE pursues the technological independence from wireless network operators, and promotes autonomy in operation and management of services.

The depicted scenarios described in this section aim to demonstrate, in a qualitative way, the potential of using ontologies for the integration of context information in management operations of services.

6.6.1 Personalized Services—CACTUS

Assume a large quantity of users that subscribe to a video conference service with QoS guarantees for image quality. This is the CACTUS service model. This application scenario, depicted in Fig. 6.8, follows the mediator's basis [Wiederhold92] for gathering raw context information of networks and users. Ontologies are used to define a policy information model to integrate user's information in the management operations, which then provides a better and advanced 3G/4G service to its users. CACTUS is responsible for providing the QoS guarantees for a specific time period, in order to hold a video conference session among the members of a group. CACTUS upgrades the services as a result of the information interoperability present in all information handling and dissemination. The CACTUS system facilitates

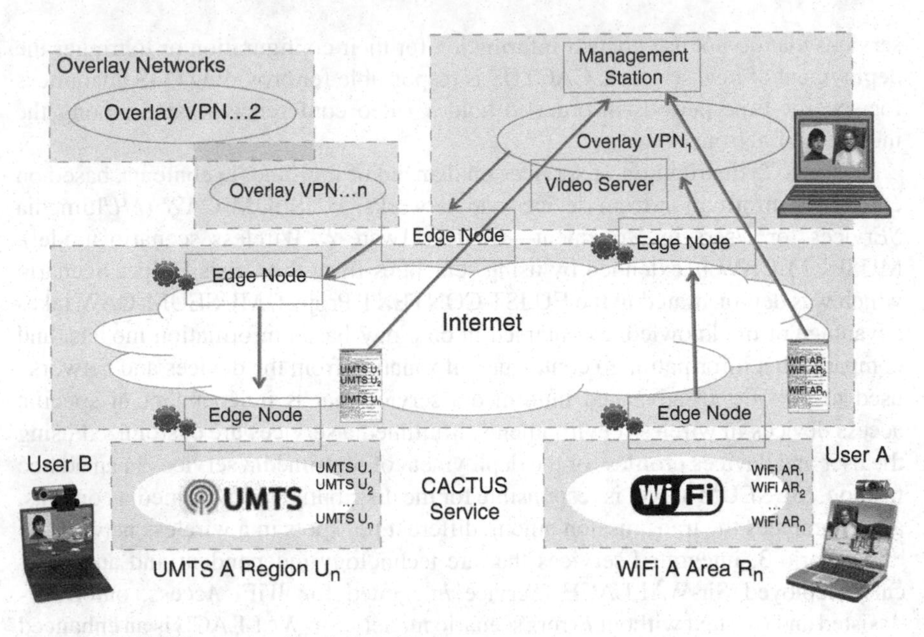

Fig. 6.8 CACTUS—depicted testbed architecture

the deployment of new services, and better manages the associated service lifecycle, as a result of the integration of different knowledge that is provided by the ontology-based information model. This enables the dynamic execution of services by using user's information to control the execution of code. The code is referred to as a service logic object (SLO).

The organizer of the conference specifies the participants and its duration, and then appropriate user profiles are generated. This provides personalized services to users as defined by the user profiles; this information is defined in ontology-based information models as end-user context information, which can be used alike as events that generate appropriate actions if some variations exist in such information. When a user registers for this service, that user enters: (a) personal information (name, address, etc.), (b) information about the network cards that he/she owns and is able to use, in order to connect to the network (e.g. MAC addresses for WLAN/LAN network cards and MSISDN numbers for UMTS/GPRS cards) and (c) the specific service level that the user chooses from among a set of available service levels. Each service level has an associated set of policies that are used to enforce the QoS guarantees for the service. The system uses this information for matching this information with the end-user profiles defined in specific ontology-based information models. Thus, if the data from an end-user profile is matched, the appropriate services are deployed and distributed and recorded in the system knowledge databases. The information contained in information model remains constant until it needs to be updated.

The conference session is scheduled by a registered user (consumer) who utilizes the conference setup service web interface to input the information for the conference session. Specifically, that user enters the conference start time, duration and

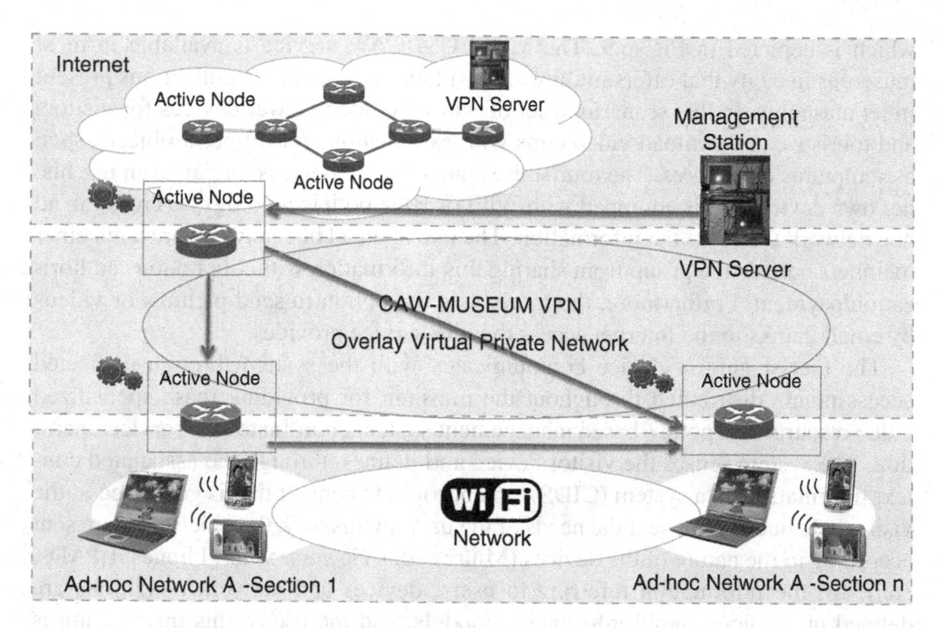

Fig. 6.9 CAW-MUSEUM—scenario architecture

the names of the participants. This information is used to customize personalized communication services, and is also categorized as an event that triggers the appropriate service based on predefined service policies. This dual use of the same information is a result of the content description at the information model and its ability to share and reuse the same information for different purposes.

Once the service's consumer schedules a conference session, a customized SLO is created, which triggers the appropriate service management policies to be generated and loaded in the policy-based management system. Then, SLOs are distributed and stored in appropriate storage points, which reflect ease of access to devices of the given topology to quickly and efficiently execute the service when the events are received. Moreover, a service invocation sensor is launched to produce the "Start Time" event when the service execution starts, which is usually when the participants connect and logon to the system. This event causes the invocation of the SLO by evaluating the context information, comparing that information with the information in the knowledge databases stored. The service is then monitored to guarantee the QoS guarantee for the image quality of the conference.

6.6.2 *Seamless Mobility Applications and Services—MUSEUM CAW*

Consider a museum-guide company offering the *MU*ltimedia *SE*rvice for *U*sers in *M*ovement, *C*ontext-*A*ware and *W*ireless service (MUSEUM-CAW scenario model), which is depicted in Fig. 6.9. The MUSEUM-CAW service is available in most museums in a city that offers multimedia explanations about the collections present in its museums. In this scenario, a set of wireless video viewer devices for visitors and tourists can download video clips with explanations for different objects, such as sculptures or pictures. The tourist uses one of these devices, or can even use his/her own device, if it is equipped with WiFi or Bluetooth technology, to create an ad hoc network to access the information. The use of an ad hoc network prevents other members outside the group from sharing this information without previous authorization/payment. Furthermore, the tourist has the option to send pictures or videos by email thanks to the Internet access that the service provides.

The tourist guide's device communicates with the system through dedicated access points distributed throughout the museum for providing these specialized video resources. A policy-based management system coordinates the service operation. The system senses the visitor device and defines, through the associated context information data system (CIDS), an appropriate context that is generated at the visitor's device to represent the needs of the user. In this way, the video clips are sent according to the nature of the devices (Multimedia Viewers, Smart Phones, I-PADs, etc.). All the information referring to users, devices and network conditions are defined in ontology-based information models, and the use of this information is based on the operation of the museum systems.

In this scenario, ontology-based information models have to be used to control the service logic automatically in order to support the automatic deployment of services. The context information dissemination system reacts to the different capabilities of the devices accessing and requiring multimedia content. In the setup of this scenario, a policy-based management system using programmable technology is required to govern the network infrastructure in which service sessions are deployed; this also governs the interaction with the ad hoc network nodes that contain the service descriptor lists for executing the SLO.

In the described scenario, WiFi and Bluetooth technology domains are dynamically linked with the CAW-MUSEUM overlay network to offer a multimedia context-aware service. Note that this is not limited to other wireless technologies; all that is required is to develop the correct brokers for processing information and wrappers for handling the raw information for other technologies.

This kind of scenario makes use of the context information contained in the ontology-based information models, for the automatic configuration of the services, which increases the functionality of the systems. It also offers other advantages, such as avoiding to have multiple devices directly connected to the museum's network.

The objective of creating ad hoc networks is to avoid the network congestion that otherwise would be produced, since a museum is visited by many people every day. Other advantage of dealing with this context information is that it allows the control of people for security purposes. The ontology-based information model contains descriptions to categorize its users, and thus identify which kind of applications must be deployed to each device. This enables an even more personalized service to be defined, since ontology-based information models are capable to define both end users and their profiles, device profiles and services for the appropriate context-dependent multimedia content and explanations. Specifically, the information model provides four categories of context: user, location, network and application context. Each one of these categories is modelled as a set of object classes defined in the ontology-based information model.

User Object. This context relates information about the user of the MUSEUM-CAW, such as the name, email address, profile and other information that uniquely describes the tourist as well as his or her personal interests. This information is then translated to a set of corresponding events that are used to trigger one or more policies that control the deployment and management of the service. Events can also be used to trigger changes to the service, as well as inform the management system of any possible network performance problems, so that the tourist's service can be protected.

Location Object. This context information defines where the user is currently located. This is important, since in this example, the ad hoc radio network operation is limited to a few meters. This is important, not just to enable the tourist to view the correct explanation for the museum piece, but also to determine if the tourist is near another group or not, and hence plan for minimizing interference. From the CAW-MUSEUM viewpoint, the system "knows" which access point the tourist is associated with, as well as the current museum piece that the tourist is viewing. This enables the system to optimize device connectivity, based on the user's current location.

Network Object. This defines the relevant information about the ad hoc network. This category includes topological and traffic data, such as access points, network nodes, bandwidth availability, capacity, network addresses, traffic levels, routing information and network security.

Application Object. This context relates the museum applications to the end users by choosing the best protocol to supply the information requested, monitoring the performance of the network and record the amount of traffic produced and bandwidth consumed during the visit.

6.6.3 Managing Complexity of Converged Networks and Services—Sir WALLACE

As an integrated management scenario, Sir WALLACE uses Policy Refinement and the PRIMO architecture to deliver its services. The main factors in this scenario are goal-oriented solutions and the non-intervention of specialized network managers. Once the service has been created, the main idea is that the user interacts with the system in a direct way, and this interaction generates pre-defined and/or customized services on demand. The network operator now has a much simpler and reduced role—to function mainly as an entity that allows and manages the operation. Only occasionally does the network operator have to intervene and define new policies to take direct control of the system, or make decisions when the system does not have enough information to process a decision on its own.

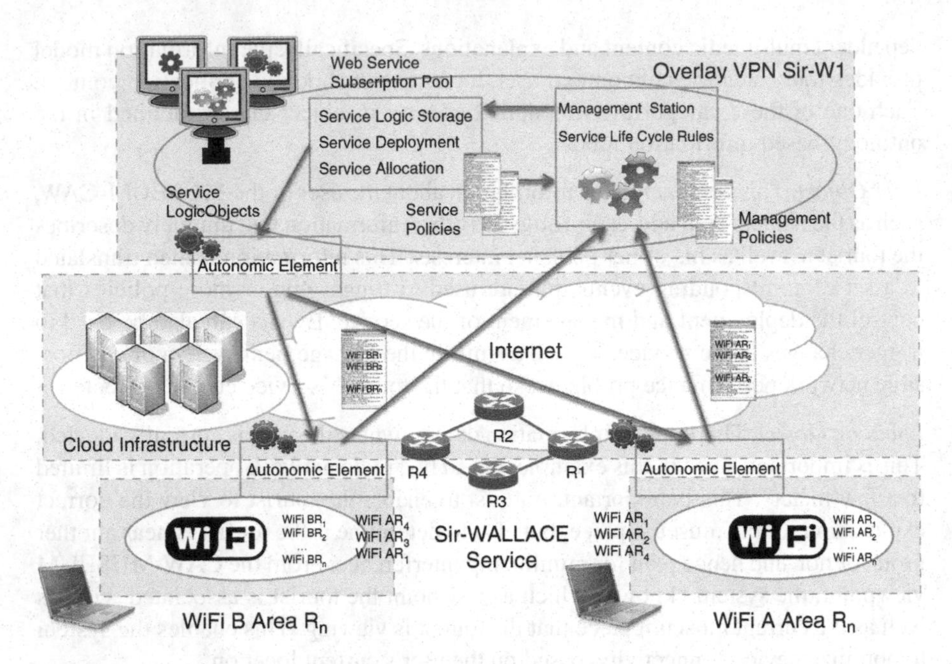

Fig. 6.10 Sir-WALLACE—depicted testbed scenario

To support management information interoperability, Sir WALLACE uses programmable components to support a Context-aware architecture [Raz99] in Onto-CONTEXT, and then provides a semantic-oriented solution using ontologies by incorporating ontology-based information models. A real next generation services scenario using the testbed of Fig. 6.10 is depicted. A precondition is that the Context platform is installed in the appropriate network edge nodes, which constitute a programmable overlay network, and that the user terminal is a mobile device that can access multiple wireless technologies, in particular IEEE 802.11b, IEEE 802.11a, and IEEE 802.11g. Specific events contained in ontology-based information models are linked with conditions and actions to create one or more policies that govern the operation. In the testbed, policies are used to trigger the evaluation of specific conditions and test some components acting as invocation service listeners according to corresponding variations of the context information in different views. Thus, context information acts as events and information.

In terms of ontology-based information models, variations in network traffic policies (such as traffic engineering) or in the network itself (such as a disconnection or link failure) is always treated as a context variation that is described in an "event" (i.e. the event is not just a simple trigger, but also contains important information, and hence is implemented as a class). Similarly, a change of wireless access technology is viewed as an "event", which represents a variation of the network status. These and other context changes are used to update functionality (e.g. connection speed, type of QoS or alternative services that can be offered), and appropriately adjust the service authentication and deployment. They can also be used to govern other business issues, such as billing, which simplifies the process for matching or mapping information between the different phases of the service lifecycle.

In the Sir-WALLACE scenario, assume that a large quantity of users all subscribe to a wireless access service, and that each is looking for independence from WiFi operators. The semantic framework approach then is used to create a service called Sir-WALLACE. The application scenario, depicted in Fig. 6.5, requires traffic engineering algorithms to satisfy the large demand for seamless mobility. Such algorithms are translated into network policies by the PRIMO architecture. The Sir-WALLACE service must also achieve interoperability between the different technologies present in order to distribute the correct information to trigger new traffic conditions within the nodes of a network to support new services.

In the scenario of Sir-WALLACE, a service takes advantage of network and user environment information, and then provides a better, more advanced wireless access service to its users using traffic engineering algorithms managed by policies defined in an ontology-based information model.

In this approach, policies are the key element using context information contained in events to modify the structure of the policy and, at the same time, collect the required information from the network. The specific traffic engineering algorithms are based on user requirements (e.g. taken from their profiles) and mobility patterns (e.g. create routing algorithms based on the most frequently used access points of the user) for the set of services that they have chosen.

The PRIMO architecture hosts the set of policies used to manage these algorithms. Sir-WALLACE is responsible for providing the QoS guarantees for a specific time period and service session; this facilitates its use among WiFi networks that use overlay networks to satisfy user demands. Sir-WALLACE upgrades the services based on the context information defined in the management policies using ontology-based information models, augmented by analyzing appropriate information obtained from the set of events received. The use of context enables the services that Sir-WALLACE provides to be better than conventional ones because the services of Sir-WALLACE use context information, which can change dynamically, to configure the appropriate service logic (SLO).

In a general manner, as part of the scenario, a user subscribes to the service (e.g. using a service setup web interface) and then, based on his or her user profile information, the system generates the appropriate services and personalizes them using the programmable nodes covering the areas in which the user is moving. These nodes are connected to the WiFi network nodes from different operators, which hide any changes in the access technology or devices used. This is an important end-user goal, as such low-level technical details should be transparent for the user.

The user information is used to infer the location of the user, and provide updated context information for building an overlay network (or VPN). However, the overlay (or VPN) is not created until the context information triggers the service. Then, the overlay (or VPN) is created and stays active while the user is present in the WALLACE system. When a user registers for this service, he enters:

(a) Personal information (name, address, etc.), which fills a profile that is modelled in information models as end-user classes; hence, this is in effect populating an object with instance data.
(b) Information about the different ways he or she can access the network, in order to connect to the network: MAC addresses for WLAN network cards, etc.; these information are also contained in the ontology-based information model as resource classes, and also can be conceptualized as populating an object with instance data.
(c) Service level, for defining QoS guarantees. A user can choose between service levels, which correspond to different policies, such as local (i.e. related to a City that is close by), region (i.e. related to a particular geographical area, which may include nearby states or countries) and global (i.e. related to many regions, such as Central and Eastern Europe). The system uses the context information to guide the future deployment of the services, and distributes the information to be stored in its appropriate knowledge databases.

6.7 Emerging Business-Driven Architectures: Ontologies and Management Systems in the Cloud

Cloud computing is the emerging technological paradigm in ICT and computing science areas. The convergence of communication resources and software mechanisms to enable multi-parallel processing (known as multi-homming) is a key factor and at the same time one of the main complex problems in cloud computing. If well is true, cloud computing has a pre-announced success because as an inherent feature cloud computing follow design pay-as-you-go bill services, up to date one of the most flexible and at the same time revolutionary economic paradigms.

Cloud computing is a result of particular features which drive the economical commits to facilitate the on-demand service provisioning and optimize server time over the Internet, reducing the billing cost. The user of services over the cloud does not own the underlying physical infrastructure or hardware devices, either pay fixed fees for dead time on servers or when not using the resources or even better no additional cost on configuration labour but rather pays for permission to use it. Whatever is the service, in today's ITC markets, business-driven challenges conduct design processes and implementation of software applications and infrastructure.

In cloud computing systems, it is true labour cost that does not disappears but rather users pay reduced prices as the cloud infrastructure is offered and used by multiple users. Literally, server capacities are shared to explode its computing processing. The processing time cost is dropped by being covered not only for one user as it happens in traditional pay for server time of use services, from here the emphasis on the concept of multitenancy in the cloud.

In the framework of cloud computing systems, new terminology has also to be adapted and the owner of the infrastructure is called cloud provider, however, it is not limited to play the role of owner, while the rent of the services is delegated to a Internet Service, provided a cloud provider also can play this role simultaneously. This fact is very difficult to find in current data centres, where the proprietary of the infrastructure is few times also the service provider. Currently, some of the main cloud providers are Amazon [AMAZON], Salesforce [SALESFORCE] and Google [APPENGINE] leading the market for their broad infrastructure and wide software portfolio of services.

Cloud computing is becoming so popular that many companies are considering, if not implement their own cloud infrastructure subsidies they service in the cloud to reduce administration, maintenance and management cost. Cloud computing is characterized by easily, administratively and technologically on-demand expansion; running of dedicated servers, providing most of the time virtual server applications according with users demands. Cloud computing offers resources when users need them and enables infrastructure offering reduced times for processing, but at the same time it allows users for closing processing sessions and as a consequence infrastructure is assigned to other users or computing purposes.

Cloud services are offered as a pay-as-you-go service and are characterized by complex pricing/economic models including time-based, utilization-based and SLA-based charges. For example, Amazon charges for an instance of information based on its size and uptime, while allowing for a reduced payment if you pre-pay for a year or 3 years for inbound and outbound network traffic, with reduced pricing with increasing capacity [AMAZON]. It also charges for disk size, reliability level, transfer rate and number of I/O operations performed.

To understand the advantage of this billing method in networking services, charges differs on locality—being free in availability zone, reduced for inter-zone and full across regions. IP addresses and load balancing are charged in addition.

In this complete new paradigm, where more complex management systems interact to exchange information the role of management systems is crucial. In one side for providing and control the system and in the other for offering and transport the information. Most of the times both activities are to be conducted in parallel, as viable alternative semantic-based systems are helping to support the multiplicity in the information, the analysis of data and the grouping of clusters of data to facilitate classification, processing and control.

The need to control multiple computers running applications and likewise the interaction of multiple service providers supporting a common service exacerbates the challenge of finding management alternatives for orchestrating between the different cloud-based systems and services. In cloud computing, a management system supporting such complex management operations must address the complex problem of coordinating multiple running applications' management operations, while prioritizing tasks for service interoperability between different cloud systems.

An emerging alternative to solve cloud computing decision control, from a management perspective is the use of formal languages as a tool for information exchange between the diverse data and information systems participating in cloud service provisioning. These formal languages rely on an inference plane [Strassner07], [Serrano09]. By using semantic decision support and enriched monitoring information, management decision support is enabled and facilitated. As a result of using semantics a more complete control of service management, operations can be offered, hence a more integrated management, which responds to business objectives. This semantically enabled decision support gives better control in the management of resources, devices, networks, systems and services, thereby promoting the management of the cloud with formal information models [Blumenthal01].

6.7.1 Services Lifecycle Control of Virtual Infrastructures—ELASTICS EMC2

As a scenario example for controlling the lifecycle of services in the cloud, ELASTICS—EMC2 (*E*lastic *S*ystems *M*anagement for *C*loud *C*omputing Infrastructures). The main factors in this scenario are goal-oriented solutions and the non-intervention of specialized network managers. In cloud computing, highly distributed and dynamic elastic virtual infrastructures are deployed in a distributed manner to support service applications. In consequence, development of management and configuration systems over virtual infrastructures are necessary. Management operations modifying the service lifecycle control loop and satisfying user demands about QoS and reliability play a critical role in this complex management process.

Cloud computing typically is characterized by large enterprise systems containing multiple virtual distributed software components that communicate across different networks and satisfying particular but secure personalized services requests [Shao10], [DEVCENTRAL]. The complex nature of these services request result in numerous data flows within the service components and the cloud infrastructures that cannot be readily correlated with each other. Ideally, data flows can be treated in run time by correlation engines, given thus the possibility to multiple cloud infrastructures and beyond boundaries the free data exchange and also cooperate to serve cloud common services.

In Fig. 6.11, the cloud service management control loop is depicted. From a data model perspective, on this control loop on-demand scalability and scalability prediction by computing data correlation between performance data models of individual components and service management operations control model is addressed. Exact component's performance modelling is very difficult to achieve since it depends on diverse number of variables. To simplify this complexity modelling on performance values (e.g. memory, CPU usage, system bus speed, and memory cache thresholds) can simplify this task. Alike to avoid complexity an estimated model calculated based on monitored data from the Data Correlation Engine to estimate the performance (represented in Fig. 6.11) can be used instead [Holub09].

An intermediate solution for efficient service control management is used, the data model used by the Service Lifecycle Rules Manager is then compared and the data linked semantically by using ontology engineering techniques. Policy-based management mechanisms are used to enable service lifecycle control operations.

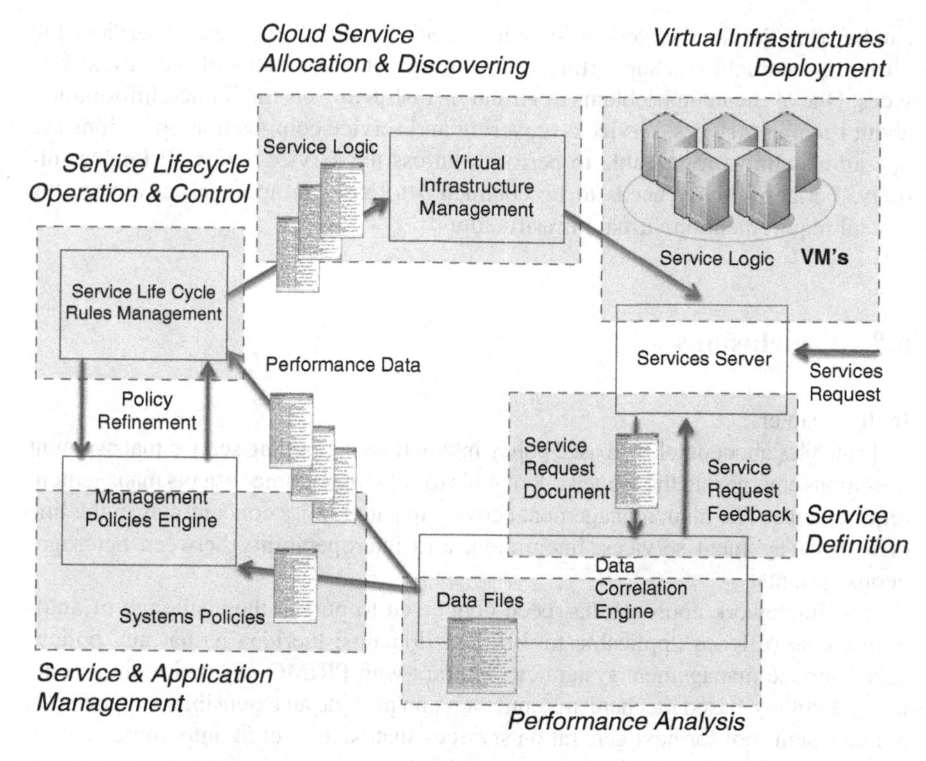

Fig. 6.11 ELASTICS EMC2—cloud infrastructure management scenario

As a result of linked data operation and information modelling an event data model is generated, which is a standardized model and can be used indistinctly in service management or network infrastructure or management domains.

Event data standard model (ontology-based) can be understood by diverse infrastructure and platforms modifying or adapting their performance according to particular applications/systems or pre-defined operation requirements. For example, Management Policies Engine can use this Event data model to define changes and dynamically perform cloud infrastructure control; this linked data standard model facilitates the dynamic adaptability of the information over infrastructures where other data models are incompatible.

The main objective of using this ontology-based linked data model is for modifying the performance of the infrastructure satisfying the general business goals according to high-level rules known as goals and defined into service contracts or service agreements. A service translation is needed in terms to define the service logic thus a virtual infrastructure manager can understand configuration instructions and execute management operations over the cloud infrastructure.

The virtual infrastructure deployment is basically the last activity in the control loop, however, variety on control of operations make this step more complex than it can be seen. Services servers must be aware all the time in the kind of services the virtual infrastructure is supporting and running, alike the status of the virtual services. One of the main problems in virtual infrastructures is the limited information about running services, service discovering and service composition operations are operations almost impossible to perform, unless the services are well known initially. Further activity needs to be conducted to have an approach covering this crucial requirement on virtual infrastructures.

6.8 Conclusions

In this chapter…

Principles about ontology-based policy interactions supporting service management operations emerge into the framework of tools to solve autonomic systems management requirements. The main management activity in autonomics concentrates in the line of self-management services, integration, and interoperability between heterogeneous systems.

The framework approach has been conceived to pursue the challenge of autonomic control loops applicable to different domains; the idea is that any policy-based service management system can interact with PRIMO and their components using ontology-based mechanisms, and then can provide an extensible and powerful management tool for next generation services that have a set of autonomic characteristics and behaviour.

The described scenarios act as examples for generic services using ontology engineering and service management mechanisms as main software and technological tools. In these scenarios, it is assumed the use of ontology-based information models and information systems. Behind the success of context-aware services, a flexible information system must exist that can accommodate heterogeneous sets of information.

The scenarios partially implemented, confirm that the desired ontology-driven functionality can be provided by using ontology-based models.

The adoption of a specific ontology-based information model is related with the nature of the service and application. According to classification studies, discussed in the state-of-the-art in this book, ontology-based modelling is accepted as the most suitable alternative in terms of composition and management costs of the information and with the advantage of application independence.

Bibliography

A

[Abowd97] Abowd, Gregory D., Dey, Anid K., Orr, R., Brotherton, J. "Context-awareness in wearable and ubiquitous computing". 1st International Symposium on Wearable Computers. p.p. 179–180, 1997.

[ACF] Autonomic Communications Forum. http://www.autonomic-communication-forum.org/og.

[ACF-FET] ACF-FET Future and Emerging Technologies Scope ftp://ftp.cordis.lu/pub/ist/docs/fet/comms-61.pdf.

[Aidarous97] Aidarous, S. and Pecyak T. (eds), "Telecomunicatios Network Management and Implementations". IEEE Press, 1997.

[Allee03] Allee, V. "The Future of Knowledge: Increasing Prosperity through Value Networks", Butterworth-Heinemann, 2003.

[AMAZON] Amazon Web Services, http://aws.amazon.com/.

[Andrews00] Andrews, Gregory R. (2000), Foundations of Multithreaded, Parallel, and Distributed Programming, Addison-Wesley, ISBN 0-201-35752-6.

[ANDROID] ANDROID Project. Active Network Distributed Open Infrastructure Development. http://www.cs.ucl.ac.uk/research/android.

[APP-PERFECT] AppPerfect DevSuite 5.0.1 – AppPerfect Java Profiler. http://www.appperfect.com/.

[APPENGINE] Google App Engine, http://code.google.com/appengine/.

[AUTOI] IST AUTOI Project, Autonomic Internet http://ist-autoi.eu.

B

[Bakker99] Bakker, J.H.L. Pattenier, F.J. "The layer network federation reference point-definition and implementation" Bell Labs. Innovation, Lucent Technol., Huizen, in TINA Conf Proc. 1999. p.p. 125–127, Oahu, HI, USA, ISBN: 0-7803-5785-X.

[Barr10] Barr, Jeff. "Host your web site in the cloud", Sitepoint, 2010, ISBN 978-0-9805768-3-2.

[Bauer03] Bauer, J. "Identification and Modelling of Contexts for Different Information Scenarios in Air Traffic", Mar. 2003. Diplomarbeit.

[Bayardo97] Bayardo, R. J. et al., "InfoSleuth: Agent-based Semantic Integration of Information in Open and Dynamic Environments". In SIGMODS, p.p. 195–206, 1997.

J.M. Serrano Orozco, *Applied Ontology Engineering in Cloud Services, Networks and Management Systems*, DOI 10.1007/978-1-4614-2236-5,

[Bearden96] Bearden, M. and Bianchini, R. "Efficient and fault-tolerant distributed host monitoring using system-level diagnosis". In Distributed Platforms: Client/Server and Beyond: DCE, CORBA, ODP and Advanced Distributed Applications, Proceedings of the IFIP/IEEE International Conference on, p.p. 159–172, 1996.

[Berners-Lee01] Berners-Lee, Tim, Hendler, James, "The semantic Web" Scientific American, May 2001.

[Bijan06] Bijan, P. et al. "Cautiously Approaching SWRL". 2006. http://www.mindswap.org/papers/CautiousSWRL.pdf.

[Blumenthal01] Blumenthal, M., Clark, D. "Rethinking the design of the Internet: the end to end arguments vs. the brave new world", ACM Transactions on Internet Technology, Vol. 1, No. 1, Aug. 2001.

[Borgida89] Borgida, A., Brachman, R., McGuinness, D., Resnick, L., "CLASSIC: A Structural Data Model for Objects", Proceedings of ACM SIGMOD conference, 1989.

[Brackman85] Brachman, R., Schmolze, J., "An Overview of the KL-ONE Knowledge Representation System", Cognitive Science, Vol. 9, No. 2, p.p. 171–216, 1985.

[Brickley03a] Brickely, D. and Miller, L. "FOAF vocabulary specification". In RDFWeb Namespace Document. RDFWeb, http://xmlns.com, 2003.

[Brickley03b] Brickley, D. and Guha, R.V. editors. "RDF Vocabulary Description Language 1.0: RDF Schema". W3C Working Draft, 2003.

[Brown00] Brown, P., Burleson, Winslow., Lamming, Mik., Rahlff, Odd-Wiking., Romano, Guy., Scholtz, Jean., Snowdon, Dave. "Context-awareness: some compelling applications", Proceedings the CHI'00 Workshop on The What, Who, Where, When, Why and How of Context-Awareness, April 2000.

[Brown98] Brown, P. J. "Triggering information by context", Personal Technologies, Vol. 2, No. 1, p.p. 1–9, September 1998.

[Brown97] Brown, P. J., Bovey, J. D., Chen, X. "Context-Aware Applications: From the laboratory to the Marketplace". IEEE Personal Communications, p.p. 58–64, 1997.

[Brown96a] Brown, M, "Supporting User Mobility". International Federation for Information Processing, 1996.

[Brown96b] Brown, P. J. "The Stick-e Document: a Framework for Creating Context-Aware Applications." Electronic Publishing '96, p.p. 259–272, 1996.

[Brown96c] Brown, P. J. "The electronic Post-it note: a model for mobile computing applications", Electronic Publishing, Vol. 9, No. 1, p.p. 1–14., September 1996.

[Brumitt00] Brumitt, B., Meyers, B., Krumm, J., Kern, A., Shafer, S., "EasyLiving: technologies for intelligent environments". Handheld and Ubiquitous Computing, September 2000.

[Brunner01] Brunner, M., Plattner, B., Stadler, R. "Service Creation and Management in Active Telecom Environments", Communications of the ACM, March 2001.

[Buyya09] Buyya, Rajkumar; Kris Bubendorfer (2009). "Market Oriented Grid and Utility Computing". Wiley. ISBN 9780470287682. http://www.wiley.com/WileyCDA/WileyTitle/productCd-0470287683,descCd-tableOfContents.html.

[Bygdås00] Bygdås, Sigrid, Malm, S. Pål, Tore, Urnes, "A Simple Architecture for Delivering Context Information to Mobile Users". Position Paper at [IFSD00], 2000.

C

[Catlett92] Catlett, Charlie; Larry Smarr (June 1992). "Metacomputing". Communications of the ACM **35** (6). http://www.acm.org/pubs/cacm/.

[CCPP] Composite Capabilities/Preference Profiles framework: http://www.w3.org/Mobile/CCPP.

[Chen04] Chen, H., Finin, T., and Joshi, A. "An Ontology for context-aware pervasive computing environments" Special issue on Ontologies for Distributed Systems, Knowledge Engineering review, 2003.

[Chen03a] Chen, H., Finin, T. and Joshi. A. "An Ontology for Context-Aware Pervasive Computing Environments". In IJCAI workshop on ontologies and distributed systems, IJCAI'03, August, 2003.

[Chen03b] Chen, H., Finin, T., and Joshi, A. "An Intelligent Broker for Context-Aware Systems". In Adjunct Proceedings of Ubicomp 2003, Seattle, Washington, USA, p.p. 12–15, October 2003.

[Chen03c] Chen, H., Finin, T., and Joshi, A. "Using OWL in a Pervasive Computing Broker". In Proceedings of Workshop on Ontologies in Open Agent Systems (AAMAS 2003), 2003.

[Chen00] Chen, G., Kotz, D., "A survey of context-aware mobile computing research", Technical Report, TR2000-381, Department of Computer Science, Dartmouth College, November 2000.

[Chen76] Chen, P. S. "The entity-relationship model: toward a unified view of data". ACM Transaction on Database Systems Vol. 1, No. 1, p.p. 9–36, March 1976.

[CHIMAERA] CHIMAERA Tool. http://www.ksl.stanford.edu/software/chimaera/.

[Clark03] Clark, D., Partridge, C., Ramming, J. C., Wroclawski, J. T. "A Knowledge Plane for the Internet". SIGCOMM 2003, Karlsruhe, Germany, 2003.

[CLEANSLATE] Clean Slate program, Stanford University, http://cleanslate.stanford.edu.

[CRICKET] CRICKET Project: http://nms.lcs.mit.edu/projects/cricket/.

[Crowcrof03] Crowcroft, J., Hand, S., Mortier, R., Roscoe, T., Warfield, A., "Plutarch: An argument for network pluralism", ACM SIGCOMM 2003 Workshops, August 2003.

D

[DAIDALOS] DAIDALOS Project: Designing Advanced network Interfaces for the Delivery and Administration of Location independent, Optimised personal Services. http://www.ist-daidalos.org/.

[DAML] Defense Agent Markup language. http://www.daml.org/.

[Damianou02] Damianou, N., Bandara, A., Sloman, M., Lupu E. "A Survey of Policy Specification Approaches", Department of Computing, Imperial College of Science Technology and Medicine, London, 2002.

[Damianou01] Damianou, N., Dulay, N., Lupu E. and Solman, M. "The Ponder Specification Language", Workshop on Policies for Distributed Systems and networks (Policy 2001). HP Labs Bristol, 29–31 January 2001.

[DARPA] DARPA Active Network Program: http://www.darpa.mil/ato/programs/activenetworks/actnet.htm.

[Davy08a] Davy, S., Jennings, B., Strassner, J. "*Efficient Policy Conflict Analysis for Autonomic Network Management*", 5th IEEE International Workshop on Engineering of Autonomic and Autonomous Systems (EASe), 2 April 2008, Belfast, Northern Ireland.

[Davy08b] Davy, S., Jennings, B., Strassner, J., "*Application Domain Independent Policy Conflict Analysis Using Information Models*", 20th Network Operations and Management Symposium (NOMS) 2008, Salvador Bahia, Brasil, 2008.

[Davy07a] Davy, S., Jennings, B., Strassner, J. "The Policy Continuum – A Formal Model", in Proc. of the 2nd International IEEE Workshop on Modelling Autonomic Communications Environments (MACE), Multlicon Lecture Notes No. 6, Multicon, Berlin, p.p. 65–78, 2007.

[Davy07b] Davy, S., Barrett, K., Jennings, B., Serrano, J.M., Strassner, J. "*Policy Interactions and Management of Traffic Engineering Services Based on Ontologies*", 5th IEEE Latin American Network Operations and Management Symposium (LANOMS), 10–12 September 2007, p.p. 95–105, ISBN 9781424411825.

[Dean02] Dean, Mike., Connolly, Dan., van Harmelen, Frank., Hendler, James., Horrocks, Ian., McGuiness, Deborah L., Patel-Schneider, Peter F., Stein, Lynn Andrea "Web Ontology Language (OWL)". W3C Working Draft 2002.

[DeBruijn04] De Bruijn, J., Fensel, D. Lara, R. Polleres, A. "OWL DL vs. OWL Flight: Conceptual Modelling and Reasoning for the Semantic Web"; November 2004.

[DeBruijn03] De Buijn, J. et al. "Using Ontologies – Enabling Knowledge Sharing and Reuse on the Semantic Web". Technical Report DERI-2003-10-29, Digital Enterprise Research Institute (DERI), Austria, October 2003.

[Debaty01] Debaty, P., Caswell, D., "Uniform Web presence architecture for people, places, and things", IEEE Personal Communications, p.p. 46–51, August 2001.

[DeVaul00] DeVaul, R.W; Pentland, A.S, "The Ektara Architecture: The Right Framework for Context-Aware Wearable and Ubiquitous Computing Applications", The Media Laboratory, MIT, 2000.

[DEVCENTRAL] The Real Meaning of Cloud Security Revealed, Online access Monday, May 04, 2009. http://devcentral.f5.com/weblogs/macvittie/archive/2009/05/04/the-real-meaning-of-cloud-security-revealed.aspx.

[Dey01] Dey, A. K., "Understanding and using context", Journal of Personal and Ubiquitous Computing, Vol. 5, No. 1, p.p. 4–7, 2001.

[Dey00a] Dey, A. K., Abowd, G. D., "Towards a better understanding of context and context awareness". In Workshop on the What, Who, Where, When and How of Context-Awareness, affiliated with the 2000 ACM Conference on Human Factors in Computer Systems (CHI 2000), April 2000, The Hague, Netherlands. April 1–6, 2000.

[Dey00b] Dey, A., K., "Providing Architectural Support for Building Context-Aware Applications", PhD thesis, Georgia Institute of Technology, 2000.

[Dey99] Dey, A.K., Salber, D., Abowd, G.D., Futakawa, M., "An architecture to support context-aware applications", GVU Technical Report Number: GIT-GVU-99-23, 1999.

[Dey98] Dey, A.K. "Context-Aware Computing: The CyberDesk Project". AAAI 1998 Spring Symposium on Intelligent Environments, Technical Report SS-98-02, p.p. 51–54, 1998.

[Dey97] Dey, A., et al. "CyberDesk: A Framework for Providing Self-Integrating Ubiquitous Software Services". Technical Report, GVU Center, Georgia Institute of Technology. GIT-GVU-97-10, May 1997.

[DMTF] Distributed Management Task Force Inc. http://www.dmtf.org.

[DMTF-CIM] DMTF, Common Information Model Standards (CIM). http://www.dmtf.org/standards/standard_cim.php.

[DMTF-DEN] DMTF, Directory Enabled Networks (DEN). http://www.dmtf.org/standards/standard_den.php.

[DMTF-DSP0005] Distributed Management Task Force, Inc. Specification for CIM Operations over HTTP. DMTF Standard DSP0005. 2003.

[DMTF-DSP0201] Distributed Management Task Force, Inc. Specification for the Representation of CIM in XML, DSP0201. 2002.

[Domingues03] Domingues, P., Silva, L., and Silva, J. "A distributed resource monitoring system". In Parallel, Distributed and Network-Based Processing, 2003. Proceedings. Eleventh Euromicro Conference on, p.p. 127–133, February 2003.

[Ducatel01] K. Ducatel, M. Bogdanowicz, F. Scapolo, J. Leijten, and J.C. Burgelman, editors. "Scenarios for Ambient Intelligence in 2010". ISTAG. 2001.

E

[Eisenhauer01] Eisenhauer, Markus and Klemke, Roland, "Contextualisation in Nomadic Computing, Ercim News", Special Issue in Ambient Intelligence, October 2001.

[Elmasri00] Elmasri, Ramez; Navathe, Shamkant B. (2000), Fundamentals of Database Systems (3rd ed.), Addison–Wesley, ISBN 0-201-54263-3.

[EU-FP7Draft] Commission of the European Communities: proposal for COUNCIL DECISIONS concerning the specific programs implementing the Framework Program 2006–2010 of the European Community for research, technological development and demonstration activities. Presented by the Commission, COM (2005), Brussels, Belgium, 2005.

[eTOM] eTOM – enhanced Telecomunication Operations Map. http://www.tmforum.org/browse.aspx?catID=1647.

F

[Feldman07] Feldmann, A. "Internet clean-slate design: what and why?," ACM SIGCOM Computer Communication Review, Vol. 37, No. 3, 2007.

[Fileto03] Fileto, R., Bauzer, C. "A Survey on Information Systems Interoperability", Technical report – IC-03-030, December 2003.

[Finkelstein01] Finkelstein, A., Savigni, A. "A Framework for Requirements Engineering for Context-Aware Services" in Proceedings of STRAW 01 the First International Workshop From Software Requirements to Architectures, 23rd International Conference on Software Engineering, 2001.

[FIPA-SC00094] Foundation for Intelligent Physical Agents. FIPA Quality of Service Ontology Specification. Geneva, Switzerland. 2002. Specification number SC00094.

[FORATV] The 'Intercloud' and the Future of Computing an Interview with Vint Cerf at FORA. tv, the Churchill Club, January 7, 2010. SRI International Building, Menlo Park, CA, Online access January 2011. http://www.fame.ie/?p=362, http://www.youtube.com/user/ForaTv#p/search/1/r2G94ImcUuY.

[Foster99] Ian, Foster; Kesselman, Carl. (1999). "The Grid: Blueprint for a New Computing Infrastructure". Morgan Kaufmann Publishers. ISBN 1-55860-475-8. http://www.mkp.com/grids/.

[Franklin98] Franklin, D., Flaschbart, J., "All Gadget and No Representation Makes Jack a Dull Environment". AAAI 1998 Spring Symposium on Intelligent Environments, Technical Report SS-98-02, p.p. 155–160, 1998.

[Fritz99] Hohl, Fritz; Kubach, Uwe; Leonhardi, Alexander; Rothermel, Kurt; Schwehm, Markus: "Next Century Challenges: Nexus – An Open Global Infrastructure for Spatial – Aware Applications", Proceedings of the Fifth Annual ACM/IEEE International Conference on Mobile Computing and Networking (MobiCom'99), Seattle, Washington, USA, T. Imielinski, M. Steenstrup, (Eds.), ACM Press, p.p. 249–255, August 15–20, 1999.

[Fritz90] Fritz E. Froehlich; Allen Kent (1990). "ARPANET, the Defense Data Network, and Internet". The Froehlich/Kent Encyclopedia of Telecommunications. 1. CRC Press. p.p. 341–375. ISBN 9780824729004. http://books.google.com/books?id=gaRBTHdUKmgC&pg=PA341.

G

[Garcia97] Garcia-Molina H., et al., "The TSIMMIS approach to mediation: Data models and Languages". Journal of Intelligent Information Systems, 1997.

[Gellersen00] Gellersen, H. W., Schmidt, A., Beigl, M. "Adding Some Smartness to Devices and Everyday Things". In the Proceedings of the Third IEEE Workshop on Mobile Computing Systems and Applications Monterey, CA, ACM, p.p. 3–10, December 2000.

[Genesereth91] Genesereth, M. "Knowledge Interchange Format" In J. Allenet & others (Eds.), 1991.

[Ghidini01] Ghidini, C., and Giunchiglia, F. "Local models semantics, or contextual reasoning locality compatibility". Artificial Intelligence Vol. 127, No. 2, p.p. 221–259, 2001.

[Gil00] Gil, Y. and Blythe, J. "PLANET: A Shareable and Reusable Ontology for Representing Plan". Proceedings of the AAAI, Workshop on Representational Issues for Real-world Planning Systems, 2000.

[Giunchiglia93] Giunchiglia, F. "Contextual reasoning. Epistemologica". Special Issue on I Linguaggi e le Macchine 16 (1993), 345–364. Also IRST-Technical Report 9211–20, IRST, Trento, Italy.

[GLITE] gLite – Lightweight Middleware for Grid Computing http://glite.cern.ch/.

[GLOBUS] Globus and Globus Toolkit. http://www.globus.org/.

[Goiri10] Goiri, I., Guitart, J. and Torres, J. "Characterizing Cloud Federation for Enhancing Providers Profit", Proceedings of IEEE 3rd International Conference on Cloud Comuting (CLOUD), p.p. 123–130, July 2010.

[GOOGLEAPP] Google app engine system status, November 2010. http://code.google.com/status/appengine.

[Gómez99] Gómez P. Asunción, Benjamins V. Richard, "Overview of Knowledge Sharing and Reuse Components: Ontologies and Problem-Solving Methods". In Proceedings of the IJCAI-99 Workshop on Ontologies and Problem-Solving Methods (KRR5), Stockholm, Sweden, 1999.

[Gray01] Gray, P., and Salber, D. "Modelling and Using Sensed Context Information in the design of Interactive Applications". In LNCS 2254: Proceedings of 8th IFIP International Conference on Engineering for Human-Computer Interaction (EHCI 2001) (Toronto/Canada, May 2001), M. R. Little and L. Nigay, Eds., Lecture Notes in Computer Science (LNCS), Springer, p. 317 ff.

[Gribble00] Gribble, Steven; Welsh, Matt; von Behren, Rob; Brewer, Eric; Culler, David; N. Borisov, S. Czerwinski, R. Gummadi, J. Hill, A. Joseph, R.H. Katz, Z.M. Mao, S. Ross, and B. Zhao. "The Ninja Architecture for Robust Internet-Scale Systems and Services". In a Special Issue of Computer Networks on Pervasive Computing. 2000.

[GRIDWAY] GridWay. http://www.gridway.org.

[Greenberg09] Greenberg, A., Hamilton, J., Maltz, D.A. and Parveen, P. "The cost of a Cloud: Research Problems in Data Center Networks" Microsoft Research, Redmon WA, USA, Editorial note submitted to CCR09.

[Gruber95] Gruber T., Towards "Principles for the Design of Ontologies Used for Knowledge Sharing", International Journal of Human-Computer studies, Vol. 43, No. 5 of 6, p.p. 907–928, 1995.

[Gruber93a] Gruber, T. R. "A translation approach to portable ontology specifications". Knowledge Acquisition, Vol. 5, No. 2, p.p. 199–220, 1993.

[Gruber93b] Gruber. T. R. "Toward principles for the design of ontologies used for knowledge sharing". Presented at the Padua workshop on Formal Ontology, March 1993.

[Gruia02] Gruia-Catalin, Roman, Christine, Julien, and Qinffeng, Huang "Network Abstractions for Context-Aware Mobile Computing", 24th International Conference on Software Engineering, Orlando (FL), ACM Press, May 2002.

[Grüninger95] Grüninger, M.S. and Fox, M. "Methodology for the Design and Evaluation of Ontologies". Procs of International Joint Conference AI Workshop on Basic Ontological Issues in Knowledge Sharing. 1995.

[Guarino95] Guarino N. & Giaretta P., "Ontologies and Knowledge Bases: Towards a Terminological Clarification, in Towards Very Large Knowledge Bases: Knowledge Building and Knowledge Sharing", N. Mars (ed.), IOS Press, Amsterdam, p.p. 25–32. 1995.

[Guerrero07] Guerrero-Casteleiro A. "Especificación del Comportamiento de Gestión de Red Mediante Ontologías". PhD Thesis, UPM, Spain, 2007.

H

[Hampson07] Hampson, C. "Semantically Holistic and Personalized Views Across Heterogeneous Information Sources", in Proceedings of the Workshop on Semantic Media Adaptation and Personalization, (SMAP07), London, UK, December 17–18, 2007.

[Harter99] Harter, A., Hopper, P., Steggles, A. and Webster, P. "The anatomy of a context-aware application" in Proceedings of MOBICOM 1999, p.p. 59–68, 1999.

[Head10] Head, M.R., Kochut, A., Shulz, C. and Shaikh, H. "Virtual Hypervisor: Enabling Fair and Economical Resource Partitioning in Cloud Environments" Proceedings of IEEE Network Operations and Management Symposium (NOMS), p.p. 104–111, 2010.

[Helin03a] Helin, H. "Supporting Nomadic Agent-based Applications in the FIPA Agent Architecture". PhD. Thesis, University of Helsinki, Department of Computer Science, Series of Publications A, No. A-2003-2. Helsinki, Finland, January 2003.

[Henricksen04] Henricksen, K., and Induska, J. "Modelling and Using Imperfect Context Information". In Workshop Proceedings of the 2nd IEEE Conference on Pervasive Computing and Communications (PerCom2004), Orlando, FL, USA, p.p. 33–37, March 2004.

[Henricksen02] Henricksen, K., Indulska, J., and Rakotonirainy, A. Modeling context information in pervasive computing systems. In LNCS 2414: Proceedings of 1st International Conference on Pervasive Computing (Zurich, Switzerland), F. Mattern and M. Naghshineh, Eds., Lecture Notes in Computer Science (LNCS), Springer, p.p. 167–180. 2002.

[Held02] Held, A., Buchholz, S., and Schill, A. "Modeling of context information for pervasive computing applications". In Proceedings of SCI 2002/ISAS 2002.

[Hightower01] Hightower, J. and Borriello, G. "Location systems for ubiquitous computing", IEEE Computer, p.p. 57–66, August 2001.

[Holub09] Holub, V., Parsons, T., O'Sullivan, P., and Murphy, J. "Run-time correlation engine for system monitoring and testing". In ICAC-INDST '09: Proceedings of the 6th international conference industry session on Autonomic computing and communications industry session, p.p. 9–18, New York, NY, USA, 2009. ACM.

[Hong01] Hong, Jason I.; Landay, James A.: An Infrastructure Approach to Context-Aware Computing. In Human-Computer Interaction, 2001, Vol. 16, 2001.

[Horn01] Horn, P., "Autonomic Computing: IBM's Perspective on the State of Information Technology", 2001.

[Horridge04] Horridge, M., Knublauch, H., Rector, A., Stevens, R., Wroe, C., "A Practical Guide to Building OWL Ontologies using the Protégé-OWL Plugin and CO-ODE Tools Edition 1.0" Uni. of Manchester, August 2004.

[Horrocks05] Horrocks, I., Parsia, B., Patel-Schneider, P. and Hendler, J. "Semantic web architecture: Stack or two towers?," in Proc. Principles and Practice of Semantic Web Reasoning (PPSWR 2005), p.p. 37–41, September 2005.

[Horrocks04] Horrocks, I., Patel-Schneider, H., Boley, H., Tabet, S., Grosof, B., and Dean, M. "SWRL: A Semantic Web Rule Language Combining OWL and RuleML" W3C Member Submission, 21 May 2004.

[Horrocks02] Horrocks, Ian. "DAM+Oil: A Reason-able Web Ontology Language". In Proceedings of the 8th Conference on Extending Database Technology (EDBT 2002), Prague, Check Republic, 2002.

[HPOPENVIEW] HP Openview event correlation services, Nov. 2010. Available [online]: http://www.managementsoftware.hp.com/products/ecs/ds/ecsds.pdf.

[Hull04] Hull, R., Kumar, B., Qutub, S.S., Unmehopa, M., Varney, D.W. "Policy enabling the services layer". Bell Labs Technical Journal, Vol. 9, No. 1, p.p. 5–18, 2004.

[Hull97] Hull, R., Neaves, P., Bedford-Roberts J. "Towards Situated Computing". 1st International Symposium on Wearable Computers, p.p. 146–153, 1997.

[Hunt98] Hunt, G.D.H., Goldszmidt, G.S., Kind, R.P., and Mukherjee, R. "Network Dispatcher: a connection router for scalable Internet services", Computer Networks and ISDN Systems, Vol. 30, p.p. 347–357, 1998.

[HYPERIC] Hyperic. Cloudstatus powered by Hyperic, November 2010. http://www.cloudstatus.com.

I

[IBM-PBM] Policy Management for Autonomic Computing. http://www.alphaworks.ibm.com/tech/pmac/.

[IBM08] IBM Software Group, U.S.A. "Breaking through the haze: understanding and leveraging cloud computing" Route 100, Somers, NY 10589. IBB0302-USEN-00. 2008.

[IBM05] IBM AC-Vision "An Architectural Blueprint for Autonomic Computing", v7, June 2005.

[IBM01a] IBM, "Autonomic Computing: IBM's Perspective on the State of Information Technology". Technical Report, IBM, 2001.

[IBM01b] IBM Autonomic Computing Manifesto. http://www.research.ibm.com/autonomic/.

[IBMTIVOLISIC] IBM. Tivoli support information center, November 2010. Available [online]: http://publib.boulder.ibm.com/tividd/td/IBMTivoliMonitoringforTransaction Performance5.3.html.

[IETF-RFC1157] J. Case et al., "A Simple Network Management Protocol (SNMP)", IETF 1157, May 1990.

[IETF-RFC2578] McCloghrie, K et al., "Structure of Management Information Version 2 (SMIv2)", IETF 2578, April 1999.

[IETF-RFC3060] Moore, E.; Elleson, J. Strassner, A. "Policy Core Information Model-Version 1 Specification". IETF Request for comments (RFC 3060), February 2001. http://www.ietf.org/rfc/rfc3060.txt.

[IETF-RFC3198] Westerinen, A.; Schnizlein, J.; Strassner, J. "Terminology for Policy-Based Management". IETF Request for Comments (RFC 3198). November 2001.

[IETF-RFC3460] Moore, E.; "Policy Core Information Model-Extensions". IETF Request for comments (RFC 3460), January 2003. http://www.ietf.org/rfc/rfc3460.txt.

[IETF-CDI] Content Distribution Interworking: http://www.content-peering.org/ietf-cdi.html.

[IETF-WI] Web Intermediaries: http://www.ietf.org/html.characters/webi-charter.html.

[IFIF] Irish Future Internet Forum. http://www.futureinternet.ie.

[IFIP-WGS] Smart Networks working group – International Federation for Information Processing IFIP. http://www.ifip.tu-graz.ac.at/TC6/WGS/index.html.

[IFIP-MNDSWG] Management of Network and Distributed Systems working group – International Federation for Information Processing IFIP. http://www.ifip.tu-graz.ac.at/TC6/WGS/index.html.

[ILOGRULES] ILOG Rules for Telecommunications. http://www.ilog.com/products/rules/.

[INMOVE] INMOVE Project "Intelligent Mobile Video Environment Project". http://www.inmove.org.

[Irfan00] Irfan A. Essa, "Ubiquitous Sensing for Smart and Aware Environments", IEEE Personal Communications, p.p. 47–49, October 2000.

[ISO8801] ISO 8801: Data Elements and Interchange Formats – Information Interchange – Representation of Date and Times – ISO, Geneva Switzerland, 2000.

[ISO11578] ISO 11578:1996: Information Technology – Open Systems Interconnection – Edition 1.

[IST-CONTEXT] IST-CONTEXT project, Active Creation, Delivery and Management of Context-Aware Services. http://context.upc.es.

[IST-EMANICS] EMANICS-European Network of Excellence for the Management of Internet Technologies and Complex Services. http://www.emanics.org/.

[ITU-X721] CCITT Recommendation X.721: Information Technology – Open Systems Interconnection – Structure of Management Information: Definition of Management Information Model – International Telecommunication Unit, Geneva, Switzerland, 1992.

[ITU-X720] CCITT Recommendation X.720: Information Technology – Open Systems Interconnection – Structure of Management Information: Management Information Model – International Telecommunication Unit, Geneva, Switzerland, 1992.

[ITU-X710] ITU-T Recommendation X.710: Information Technology – Open Systems Interconnection – Common Management Information Service – International Telecommunication Unit, Geneva, Switzerland, 1997.

[ITU-X701] ITU-T Recommendation X.701: Information Technology – Open Systems Interconnection – Systems Management Overview – International Telecommunication Unit, Geneva, Switzerland, 1997.

[ITU-X700] ITU-T Recommendation X.700: Management Framework for Open Systems Interconnection – Open Systems Interconnection for CCITT Applications – International Telecommunication Unit, Geneva, Switzerland, September 1992.

J

[Jeng03] Jeng, Jun-Jang., Chang, H. and Chung, Jen-Yao. "A Policy Framework for Business Activity Management". E-Commerce, IEEE International Conference. June 2003.

[Joshi03] Joshi, A. "A Policy Language for a Pervasive Computing Environment". In Procceedings of IEEE 4th International Workshop on policies for Distributed Systems and Networks, 2003. POLICY 2003.

K

[Kagal03] Kagal, L., Finin, T. and Joshi, A. "A Policy-Based Approach to Security for the Semantic Web," Proceedings. 2nd Int'l Semantic Web Conf. (ISWC 2003), LNCS 2870, Springer-Verlag, 2003, p.p. 402–418.

[Kagal02] Kagal, L. "REI: A Policy Language for the Me-Centric Project" HP Labs, Technical Report hpl-2002-070, September 2002.

[Kanter02] Kanter, T. G., "Hottown, enabling context-Aware and extensible mobile interactive spaces", Special Issue of IEEE Wireless Communications and IEEE Pervasive on "Context-Aware Pervasive Computing", p.p. 18–27, October 2002.

[Kanter00] Kanter, T., Lindtorp, P., Olrog, C., Maguire, G.Q., "Smart delivery of multimedia content for wireless applications", Mobile and Wireless Communication Networks, p.p. 70–81, 2000.

[Kantar03] Kantar, T.G., Gerald Q. Maguire Jr., Smith, M. T., "Rethinking Wireless Internet with Smart Media', 2003 http://psi.verkstad.net/Papers/conferences/nrs01/nrs01-theo.PDF.

[Karmouch04] Karmouch, A., Galis, A., Giaffreda, R., Kanter, T., Jonsson, A., Karlsson, A. M. Glitho, R. Smirnov, M. Kleis, M. Reichert, C., Tan, A., Khedr, M., Samaan, N., Heimo, L., Barachi, M. E., Dang, J. "Contextware Research Challenges in Ambient Networks" ISBN 3-540-23423-3, Springer-Verlag Lecture Notes in Computer Science-IEEE MATA 2004, Florianopolis, Brazil, 20–22 October 2004.

[Katsiri05] Katsiri, E., "Middleware Support for Context-Awareness in Distributed Sensor-Driven Systems". PhD Thesis, Engineering Department, University of Cambridge. Also published as Technical Report n.620, Computer Laboratory, University of Cambridge, February 2005.

[Keeney06] Keeney, J., Lewis, D., O'Sullivan, D., Roelens, A., Boran, A. "Runtime Semantic Interoperability for Gathering Ontology-based Network Context", 10th IEEE/IFIP Network Operations and Management Symposium (NOMS 2006), Vancouver, Canada, p.p. 56–65, April 2006.

[Keeney05] Kenney, J., Carey, K., Lewis, D., O'Sullivan, D., Wade, V. "Ontology-based Semantics for Composable Autonomic Elements". Workshop on AI in Autonomic Communications, 19th International Joint Conference on Artificial Intelligence, Edinburgh, Scotland, 30 July–5th August 2005.

[Kephart03] Kephart, J. O. and Chess, D. M., "The Vision of Autonomic Computing", IEEE Computer Magazine, January 2003. http://research.ibm.com/autonomic/research/papers/.

[Khedr03] Khedr, M. and Karmouch, A. "Exploiting SIP and agents for smart context level agreements", 2003 IEEE Pacific Rim Conference on Communications, Computers and Signal Processing, Victoria, BC, Canada, August 2003.

[Khedr02] Khedr, M., Karmouch, A., Liscano, R. and Gray, T. "Agent-based context aware ad hoc communication". In Proceedings of the 4th International Workshop on Mobile Agents for Telecommunication Applications (MATA 2002), Barcelona, Spain, p.p. 292–301, Oct 23–24, 2002.

[KIF] KIF Language – Knowledge Interchange Format Language. http://www-ksl.stanford.edu/knowledge-sharing/kif/.

[Kitamura01] Kitamura, Y., Kasai, T., and Mizoguchi,R. "Ontology-based Description of Functional Design Knowledge and its Use in a Functional Way Server". Proceedings of the Pacific Conference on Intelligent Systems 2001, 2001.

[Kirk95] Kirk, T., Levy, A., Sagiv, Y. and Srivastava, D. "The Information Manifold", AAAI Spring Symposium on Information Gathering, 1995.

[Kobielus06] Kobielus, J. "New Federation Frontiers In IP Network Services", Publication: Business Communications Review. Date: Tuesday, August 1 2006.

[Komblum00] Kornblum, Jessica. Raz, Danny, Shavitt, Yuval. "The Active Process Interaction with its Environment" Computer and Information Science, Univ. of Pennsylvania, Bell Labs, Lucent Technologies. Holmdel, NJ. 2000.

[Korpiää03b] Korpipää, P. Mäntyjärvi, J., Kela, J., Keränen, H. and Malm. E-J. "Managing Context Information in Mobile Devices". IEEE Pervasive Computing Vol. 2, No. 3, p.p. 42–51. 2003.

[Kouadri04] Kouadri Mostefaoui, G. and Brezillon, P. "Modelling Context-Based Security Policies with Contextual Graphs". In Workshop on Context Modeling and Reasoning, 2004.

[Klemke00] Klemke, Roland. "Context Framework – an Open Approach to Enhance Organisational Memory Systems with Context Modelling Techniques', in PAKM-00: Practical Aspects of Knowledge Management, Proc. 3rd International Conference, Basel. Switzerland, 2000.

[Klemke01] Klemke, Roland., Nick, Achim. "Case Studies in Developing Contextualising Information Systems", in: CONTEXT-01 – Third International and Interdisciplinary Conference on Modeling and Using Context, Dundee (Scotland), July 27–30, 2001.

[Korke-Aho00] Korkea-aho, Mari. "Context-Aware Applications Survey", Internetworking Seminar (Tik-110.551), Helsinki University of Technology. Spring 2000.

[Krause05] Krause, M., Hochstatter, I., "Challenges in Modeling and Using Quality of Context", ISBN10 3-540-29410-4, Springer-Verlag Lecture Notes in Computer Science-IEEE MATA 2005, p.p. 17–19, Montreal, Canada, October 2005.

L

[Lewis06] Lewis, D., O'Sullivan, D., Feeney, K., Keeney, J., Power, R. "Ontology-based Engineering for Self-Managing Communications", 1st IEEE International Workshop on Modelling Autonomic Communications Environments (MACE 2006), Dublin, Ireland, 25–26 October 2006, edited by W. Donnelly, R. Popescu-Zeletin, J. Strassner, B. Jennings, S. van der Meer, multicon verlag, p.p. 81–100, 2006.

[Liao07] Liao, L., Leung, H.K.N. "An Ontology-based Business Process Modeling Methodology", in Advances in Computer Science and Technology - ACST 2007. 2–4 April 2007, Phuket, Thailand.

[Long96] Long, Sue., Kooper, Rob., Abowd, Gregory D. and Aktenson, Christopher G., "Rapid Prototyping of Mobile Context-Aware Applications: The Cyberguide Case Study". Proceedings of the second annual international conference on Mobile computing and networking, p.p. 97–107, Rye, New York, United States, November 1996.

[López03a] López de Vergara, Jorge E., Villagrá, Víctor A., Berrocal, Julio, Asensio, Juan I., "Semantic Management: Application of Ontologies for the Integration of Management Information Models". In Proceedings of the 8th IFIP/IEEE International Symposium on Integrated Management (IM 2003), Colorado Springs, Colorado, USA, 2003.

[López03b] López de Vergara, Jorge E., Villagrá, Víctor A., Asensio, Juan I., Berrocal, Julio., "Ontologies: Giving Semantic to Network Management Models". IEEE Network Magazine, Special Issue on network Management, Vol. 17, No. 3, May 2003.

[López03c] López de Vergara, J.E. "Especificación de modelos de información de gestión de red integrada mediante el uso de ontologías y técnicas de representación del conocimiento". PhD Thesis, UPM, Spain, 2003.

[LOVEUS] LOVEUS Project. "Location Aware Visually Enhanced Ubiquitous Services Project": http://loveus.intranet.gr.

[Lynch96] Lynch, Nancy A. (1996), Distributed Algorithms, Morgan Kaufmann, ISBN 1-55860-348-4.

M

[Mace11] Mace, J.C., van Moorsel, A. and Watson, P."The case of dynamic security solutions in public cloud workflow deployments" Proceedings of IEEE/IFIP 41st International Conference on Dependable Systems and networks Workshops (DNS-W), p.p. 111–116, June 2011.

[Maozhen05] Maozhen, Li., Baker, Mark, A. (2005). "The Grid: Core Technologies". Wiley. ISBN 0-470-09417-6. http://coregridtechnologies.org/.

[McCarthy97] McCarthy, J., and Buva Cˇ. "Formalizing context (expanded notes)". In Working Papers of the AAAI Fall Symposium on Context in Knowledge Representation and Natural Language (Menlo Park, California, 1997, American Association for Artificial Intelligence, p.p. 99–135. 1997.

[McCarthy93] McCarthy, J. "Notes on formalizing contexts". In Proceedings of the Thirteenth International Joint Conference on Artificial Intelligence, San Mateo, California, 1993, R. Bajcsy, Ed., Morgan Kaufmann, p.p. 555–560. 1993.

[McGuiness02] McGuiness, L. Deborah., Fikes, Richard., Hendler, James., Lynn Andrea, "DAM+OIL: An Ontology Language for the Semantic Web", in IEEE Intelligent Systems, Vol. 17, No. 5 September 2002.

[Mei06] Mei, J., Boley, H. "Interpreting SWRL Rules in RDF Graphs". Electronic Notes in Theoretical Computer Science (Elsevier) (151): 53–69. 2006.

[MicrosoftPress11] The Economics of the cloud, online access Wednesday 05, January 2011. http://www.microsoft.com/presspass/presskits/cloud/docs/The-Economics-of-the-Cloud.pdf.

[Mitra00] Mitra, P., Wiederhold, G., Kersten, M. "A graph-oriented model for articulation of Ontology Interdependencies", In Proceedings of the Conference on Extending Database Technology 2000 (EDBT 2000) Konstanz, Germany, March 2000.

N

[Nakamura00] Nakamura, Tetsuya., Nakamura, Matsuo., Tomoko, Itao. "Context Handling Architecture for Adaptive Networking Services". Proceedings of the IST Mobile Summit 2000.

[Neches91] Neches, Robert., Fikes, Richard., Finin, Tim., Patil, Ramesh., Senator, Ted. Swartout, William, R. "Enabling Technology for Knowledge Sharing". AI Magazine, Vol. 12, No. 3, 1991.

[NEWARCH] Clark, D et al., "NewArch: Future Generation Internet Architecture", NewArch Final Technical Report, http://www.isi.edu/newarch/.

[NGN] Architecture Design Project for New Generation Network, http://akari-project.nict.go.jp/eng/index2.htm.

[Neiger06] Neiger, G., Santoni, A., Leung, F., Rodgers, D. and Uhlig, R. "Intel Virtualization Technology: Software-only virtualization with the IA-32 and Itanium architectures", Intel Technology Journal, Vol. 10, No. 03, August 2006.

[Novak07] Novak, J.: "Helping Knowledge Cross Boundaries: Using Knowledge Visualization to Support Cross-Community Sensemaking", in Proceedings of the Conference on System Sciences, (HICSS-40), Hawaii, January 2007.

[NSFFI] NSF-funded initiative to rebuild the Internet, http://www.geni.net/.

O

[Ocampo05c] Ocampo, R., Cheng, L., and Galis, A., "ContextWare Support for Network and Service Composition and Self-Adaptation". IEEE MATA 2005, Mobility Aware Technologies and Applications, Service Delivery Platforms for Next Generation Networks; Springer ISBN-2 553-01401-5, p.p. 17–19, Montreal, Canada, October 2005.

[OKBC] Open Knowledge Base Connectivity language – Specification. http://www.ai.sri.com/~okbc/spec.html.

[OMG-MDA] Object Management Group. Model Driven Architecture. http://www.omg.org/mda/.

[OMG-UML] Object Management Group, "Unified Modelling Language (UML), version 1.4, UML Summary", OMG document, September 2001.

[ONTOLINGUA] ONTOLINGUA Description Tool. http://www.ksl.stanford.edu/software/ontolingua.

[OPENDS] OpenDS Monitoring. https://www.opends.org.

[OPES] Open Pluggable Edge Services – OPES: http://www.ietf-opes.org.

[Opperman00] Oppermann, Reinhard; Specht, Marcus. "A Context-sensitive Nomadic Information System as an Exhibition Guide". Handheld and Ubiquitous Computing Second International Symposium. 2000.

[O'Sullivan03] O'Sullivan, D., Lewis, D. "Semantically Driven Service Interoperability for Pervasive Computing". In Proceedings of 3rd ACM International Workshop on Data Engineering for Wireless and Mobile Access, San Diego, California, USA, September 19th, 2003.

[OWL] Ontology Web Language, http://www.w3.org/2004/OWL.

[OWL-S] http://www.daml.org/services/owl-s/.

P

[Palpanas07] Palpanas, T., Chowdhary, P., Mihaila, G.A., Pinel, F.: "Integrated model-driven dashboard development", in the Journal of Information Systems Frontiers, Vol. 9, No. 2–3, Jul 2007.

[Park04] Park, Jinsoo., Ram, Sudha. "Information systems interoperability: What lies beneath?". ACM Transactions on Information Systems 22–4, 2004.

[Perich04] F. Perich. "MoGATU BDI Ontology", University of Maryland, Baltimore County 2004.

[Pascoe99] Pascoe, J., Ryan, N., Morse, D., "Issues in developing context-aware computing". In Proceedings of First International Symposium on Handheld and Ubiquitous Computing (HUC'99), 1999.

[Pascoe98] Pascoe, J. "Adding Generic Contextual Capabilities to Wearable Computers". 2nd International Symposium on Wearable Computers, p.p. 92–99, 1998.

[Piccinelli01] Piccinelli, Giacomo., Stefanelli, Cesare. Morciniec, Michal. "Policy-based Management for E-Services Delivery" HP-OVUA 2001.

[Plazczak06] Plaszczak, Pawel; Rich Wellner, Jr (2006). "Grid Computing The Savvy Manager's Guide". Morgan Kaufmann Publishers. ISBN 0-12-742503-9. http://savvygrid.com/.

[PROMPT] PROMPT Tool http://protege.stanford.edu/plugins/prompt/prompt.html.

[PROTÉGÉ] PROTÉGÉ http://protege.stanford.edu/.

R

[Raz99] Raz, D. and Shavitt, Y. "An Active Network Approach for Efficient Network Management", IWAN'99, Berlin, Germany, LNCS 1653, p.p. 220–231, July 1999.

[Reynaud03] Reynaud, C., Giraldo, G., "An application to the mediator approach to services over the web", in Concurrent Engineering, 2003.

[RDF] http://www.w3c.org/rdf.

[Rochwerger11] Rochwerger, B. et. Al., "Reservoir When One Cloud Is Not Enough". Computer Magazine, Vol. 44, No. 3, p.p. 44–51, March 2011.

[Rochwerger09] Rochwerger, B., Caceres, J., Montero, R.S., Breitgand, D., Elmroth, E., Galis, A., Levy, E. Llorente, I.M., Nagin, K., Wolfsthal, Y., Elmroth, E., Caceres, J., Ben-Yehuda, M., Emmerich, W., Galan, F. "The RESERVOIR Model and Architecture for Open Federated Cloud Computing", IBM Journal of Research and Development, Vol. 53, No. 4. 2009.

[Roussaki06] Roussaki, I. M., Strimpakou, M., Kalatzis, N., Anagnostou, M. and Pils, C. "Hybrid Context Modeling: A location-based scheme using ontologies". In 4th IEEE conference on Pervasive Computing and Communications Workshop, p.p. 2–7, 2006.

[Ryan97] Ryan, N., Pascoe, J., Morse, D. "Enhanced Reality Fieldwork: the Context-Aware Archaeological Assistant". Gaffney, V., van Leusen, M., Exxon, S. (eds.) Computer Applications in Archaeology, 1997.

S

[Salber99] Salber, D, Dey, A.K., Abowd, G.D., The Context Toolkit: Aiding the Development of Context-Enabled Applications, in Proceedings of CHI'99, PA, ACM Press, p.p. 434–441, May 1999.

[SALESFORCE] Salesforce.com, http://www.salesforce.com/cloudcomputing/.

[Samann03] Samann, N., Karmouch, A., "An Evidence-based Mobility Prediction Agent Architecture". In Proceedings of the 5th Int. Workshop on Mobile Agents for Telecommunication Applications (MATA2003), Marrakesch, ISBN 3-540-20298-6, Lecture Notes in Computer Science, Springer-Verlag, October 2003.

[Schilit95] Schilit, B.N. (1995) "A Context-Aware System Architecture for Mobile Distributed Computing" PhD Thesis 1995.

[Schilit94a] Schilit, B; Theimer, M. "Disseminating Active Map Information to Mobile Hosts". IEEE Network Vol. 8, No. 5, p.p. 22–32, 1994.

[Schilit94b] Schilit, B. N., Adams, N. L., and Want, R. "Context-aware computing applications". In IEEE Workshop on Mobile Computing Systems and Applications, Santa Cruz, CA, USA, p.p. 85–90, 1994.

[Schönwälder99] Schönwälder, J. Straub F. "Next Generation Structure of Management Information for the Internet". In Proceedings of the 10th IFIP/IEEE International Workshop on Distributed Systems: Operations & Management (DSOM'99), Zürich, 1999.

[Schmidt02] Schmidt, A., Strohbach, M., van Laerhoven, K., Friday A. and Gellersen, H.W. "Context Acquisition based on Load Sensing", in Proceedings of Ubicomp 2002, G. Boriello and L.E. Holmquist (Eds). Lecture Notes in Computer Science, Vol. 2498, ISBN 3-540-44267-7; Göteborg, Sweden. Springer Verlag, p.p. 333–351, September 2002.

[Schmidt01] Schmidt, A., van Laerhoven, K., "How to Build Smart Appliances?", IEEE Personal Communications, Vol. 8, No. 4, August 2001.

[Sedaghat11] Sedaghat, M., Hernandez, F. And Elmroth, E. "Unifying Cloud Management: Towards Overall Governance of Business Level Objectives", Proceedings of 11th IEEE/ACM International Symposium on Cluster, Cloud and Grid Computing (CCGrid), p.p. 591–597, May 2011.

[Serrano10] Serrano J.M., Van deer Meer, S., Holum, V., Murphy J., and Strassner J. "Federation, A Matter of Autonomic Management in the Future internet". 2010 IEEE/IFIP Network Operations & Management Symposium – NOMS 2010. Osaka International Convention Center, 19–23 April 2010, Osaka, Canada.

[Serrano09] Serrano, J.M., Strassner, J. and ÓFoghlú, M. "A Formal Approach for the Inference Plane Supporting Integrated Management Tasks in the Future Internet" 1st IFIP/IEEE ManFI International Workshop, In conjunction with 11th IFIP/IEEE IM2009, 1–5 June 2009, at Long Island, NY, USA.

[Serrano08] Serrano, J. M., Serrat, J., Strassner, J., Ó Foghlú, Mícheál. "Facilitating Autonomic Management for Service Provisioning using Ontology-Based Functions & Semantic Control" 3rd IEEE International Workshop on Broadband Convergence Networks (BcN 2008) in IEEE/IFIP NOMS 2008. 07–11 April 2008, Salvador de Bahia, Brazil.

[Serrano07a] Serrano, J. Martín; Serrat, Joan; Strassner, John, "Ontology-Based Reasoning for Supporting Context-Aware Services on Autonomic Networks" 2007 IEEE/ICC International Conference on Communications – ICC 2007, 24–28 June 2007, Glasgow, Scotland, UK.

[Serrano07b] Serrano, J. Martín; Serrat, Joan, Meer, Sven van der, Ó Foghlú, Mícheál "Ontology-Based Management for Context Integration in Pervasive Services Operations". 2007 ACM International Conference on Autonomous Infrastructure Management and Security. AIMS 2007, 21–23 June 2007, Oslo, Norway.

[Serrano07c] Serrano, J. M., Serrat, J., Strassner, J., Cox, G., Carroll, R., Ó Foghlú, M. "Services Management Using Context Information Ontologies and the Policy-Based Management Paradigm: Towards the Integrated Management in Autonomic Communications". 2007 1st IEEE Intl. Workshop on Autonomic Communications and Network Management – ACNM 2007, in 10th IFIP/IEEE International Symposium on Integrated Management – IM 2007, 21–25 May 2007, Munich, Germany.

[Serrano06a] Serrano, J. Martín, Serrat, Joan, Strassner, John, Carroll, Ray. "Policy-Based Management and Context Modelling Contributions for Supporting Autonomic Systems". IFIP/TC6 Autonomic Networking. France Télécom, Paris, France. 2006.

[Serrano06b] Serrano, J.M., Serrat, J., O'Sullivan, D. "Onto-Context Manager Elements Supporting Autonomic Systems: Basis & Approach". IEEE 1st Int. Workshop on Modelling Autonomic Communications Environments – MACE 2006. Manweek 2006, Dublin, Ireland. October 23–27, 2006.

[Serrano06c] Serrano, J.M., Justo, J., Marín, R., Serrat, J., Vardalachos, N., Jean, K., Galis, A. "Framework for Managing Context-Aware Multimedia Services in Pervasive Environments". 2006 International Journal on Internet Protocol and Technologies – IJIPT Journal, Special Issue on Context in Autonomic Communication and Computing. Vol. 1, 2006. ISSN 1743–8209 (Print), ISSN 1743–8217 (On-Line).

[Serrano06d] Serrano, J.M., Serrat, J., Galis, A. "Ontology-Based Context Information Modelling for Managing Pervasive Applications". 2006 IEEE/IARIA International Conference on Autonomic and Autonomous Systems – ICAS'06. AWARE 2006. July 19–21, 2006 – Silicon Valley, CA, USA.

[Serrano05] Jaime M. Serrano O., Joan Serrat F., Kun Yang, Epi Salamanca C. "Modelling Context Information for Managing Pervasive Network Services". 2005 AMSE/IEEE International Conference on Modelling and Simulation – ICMS' 05, AMSE/IEEE Morocco Section, 22–24 November 2005, Marrakech, Morocco.

[SFIFAME] SFI FAME-SRC, Federated, Autonomic Management of End-to-End Services – Strategic Research Cluster. http://www.fame.ie/.

[Shao10] Shao, J., Wei, H., Wang, Q. and Mei, H. "A runtime model based monitoring approach for cloud". In Cloud Computing (CLOUD), 2010 IEEE 3rd International Conference on, p.p. 313–320, July 2010.

[Sloman94a] Sloman, M. (ed.), "Network and distributed systems management", Addison-Wesley, 1994.

[Sloman94b] M. Sloman. "Policy Driven Management for Distributed Systems". Journal of Network and Systems Management, p.p. 215–333, 1994.

[Sloman99c] M. Sloman, and E. Lupu, "Policy Specification for Programmable Networks", International Working Conference on Active Networks (IWAN'99), Berlin, Germany, Springer-Verlag LNCS, June–July 1999.

[Srikanth09] Srikanth Kandula, "The Nature of Datacenter Traffic: Measurements & Analysis." IMC'09, November 4–6, 2009, Chicago, Illinois, USA. IMC '09 Proceedings of the 9th ACM SIGCOMM conference on Internet measurement conference. ACM New York, NY, USA 2009 ISBN: 978-1-60558-771-4.

[Starner98] Starner, T., Kirsh, D., Pentland, A. "Visual context awareness in wearable computing". In Digest of Papers. 2nd International Symposium on Wearable Computers, p.p. 50–57, 1998.

[Strang04] Strang, T., Linnhoff-Popien, C., "A Context Modelling Survey". Workshop on Advanced Context Modelling, Reasoning and Management Nottingham, England; September 2004.

[Strang03b] Strang, T., Linnhoff-popien, C., and Frank, K. "Applications of a Context Ontology Language". In Proceedings of International Conference on Software, Telecommunications and Computer Networks (SoftCom2003), University of Split, Croatia, p.p. 14–18. October 2003.

[Strang03c] Strang, T., Linnhoff-Popien, C., and Frank, K. "CoOL: A Context Ontology Language to enable Contextual Interoperability". In LNCS 2893: Proceedings of 4th IFIP WG 6.1 International Conference on Distributed Applications and Interoperable Systems (DAIS2003), Springer Verlag, p.p. 236–247, Paris/France, November 2003.

[Strassner08] Strassner, J., Raymer, D., Samudrala, S., Cox, G., Liu, Y., Jiang, M., Zhang, J., van der Meer, S., Jennings, B., Ó Foghlú, M., Donnelly, W. "The Design of a New Context-Aware Policy Model for Autonomic Networking". Proc. of 5th IEEE International Conference on Autonomic Computing (ICAC 2008) Chicago, Illinois, USA, June 2–6, 2008.

[Strassner07b] Strassner, J., Ó Foghlú, M., Donnelly, W. Agoulmine, N. "Beyond the Knowledge Plane: An Inference Plane to Support the Next Generation Internet", IEEE GIIS 2007, 2–6 July, 2007.

[Strassner07a] Strassner, J., Neuman de Souza, J., Raymer, D., Samudrala, S., Davy, S., Barret, K. "The Design of a New Policy Model to Support Ontology-Driven Reasoning for Autonomic Networking", 5th Latin American Network Operations and Management Symposium – LANOMS 2007, LNCC – Petrópolis, Brazil, September 10–12, 2007.

[Strassner07] Strassner, J., "Knowledge Engineering Using Ontologies", chapter in Elsevier Handbook of Network and System Administration. Jan Bergstra and mark Burgues Eds. Elsevier, 2007.

[Strassner06a] Strassner, J. and Kephart, J., "Autonomic Networks and Systems: Theory and Practice", NOMS 2006 Tutorial, April 2006.

[Strassner06b] Strassner, J., "Seamless Mobility – A Compelling Blend of Ubiquitous Computing and Autonomic Computing", in Dagstuhl Workshop on Autonomic Networking, January 2006.

[Strassner06c] Strassner, J., Lehtihet, E., Agoulmine, N., "FOCALE – A Novel Autonomic Computing Architecture", LAACS 2006.

[Strassner04] Strassner, J., "Policy Based Network Management", Morgan Kaufmann, ISBN 1-55860-859-1, 2004.

[Strassner02] Strassner, J., "DEN-ng: achieving business-driven network management", Network Operations and Management Symposium (NOMS 2002), ISBN 1-55860-859-1, 2002.

[Sure02] Sure, Y., Staab, S. and Studer, R. "Methodology for Development and Employment of Ontology based Knowledge Management Applications" SIGMOD Rec., 31(4), 2002.

[Swartout96] Swartout, Bill., Patil, Ramesh., Knight, Kevin and Russ, Tom. "Toward Distributed Use of Large-Scale Ontologies" In Proceedings of the Tenth Knowledge Acquisition for Knowledge-Based Systems Workshop, Banff, Alberta, Canada. November 9–14, 1996.

T

[TMF-ADDENDUM] TMF, "The Shared Information and Data Model – Common Business Entity Definitions: Policy", GB922 Addendum 1-POL, July 2003.

[TMF-SID] SID – Shared Information Data model. http://www.tmforum.org/InformationManagement/1684/home.html.

[TMN-M3010] Telecommunications Management Networks – Architectural Basis. http://www.simpleweb.org/tutorials/tmn/index-1.html#recommendations.

[TMN-M3050] Telecommunications Management Networks – Management Services approach – Enhanced Telecommunications Operations Map (eTOM). http://www.catr.cn/cttlcds/itu/itut/product/bodyM.htm.

[TMN-M3060] Telecommunications Management Networks – Principles for the Management of Next Generation Networks. http://www.catr.cn/cttlcds/itu/itut/product/bodyM.htm.

[Tennenhouse97] Tennenhouse, D.L., Smith, J. M., W. D. Sincoskie, D. J. Wetherall, and G. J. Minden, (1997) "A Survey of Active Network Research," IEEE Communications, Vol. 35, No. 1, p.p. 80–86, January 1997.

[TERAGRID] TeraGrid. https://www.teragrid.org.

[Tomlinson00] Tomlinson, G., Chen, R., Hoffman, M., Penno, R. "A Model for Open Pluggable Edge Services", draft-tomlinson-opes-model-00.txt, http://www.ietf-opes.org.

[Tonti03] Tonti, G., Bradshaw, R., Jeffers, R., Suri, N. and Uszok, A. "Semantic Web Languages for Policy Representation and Reasoning: A Comparison of KAoS, Rei, and Ponder," The Semantic Web–ISWC 2003: 2nd International Semantic Web Conference, LNCS 2870, Springer-Verlag, 2003, p.p. 419–437.

U

[UAPS] User Agent Profile specification: http://www.openmobilealliance.org.

[UNICORE] UNICORE (Uniform Interface to Computing Resources. http://www.unicore.eu/.

[Urgaonkar10] Urgaonkar, R., Kozat, U.C., Igarashi, K. and Neely, M.J. "Dynamic Resource Allocation and Power Management in Virtualized Data Centers" Proceedings of IEEE Network Operations and Management Symposium (NOMS), p.p. 479–486, April 2010.

[Uschold96] Ushold M. & Gruninger M., "Ontologies: Principles, methods and applications, in The Knowledge Engineering Review", Vol. 11, No. 2, p.p. 93–155, 1996.

[Uszok04] Uszok,A. Bradshaw, J.M. and Jeffers, R. "KAoS: A Policy and Domain Services Framework for Grid Computing and Grid Computing and Semantic Web Services," Trust Management: 2nd Int'l. Conference Procs (iTrust 2004), LNCS 2995, Springer-Verlag, 2004, p.p. 16–26.

V

[Verma00] Verma D. "Policy Based Networking" 1st ed. New Riders. ISBN: 1-57870-226-7 Macmillan Technical Publishing USA, 2000.

[VMWARE] Cisco, VMWare. "DMZ Virtualization using VMware vSphere 4 and the Cisco Nexus" 2009. http://www.vmware.com/files/pdf/dmz-vsphere-nexus-wp.pdf.

W

[W3C] World Wide Web Consortium (3WC). http://www.w3.org.

[W3C-WebServices] W3C Consortium – WebServices Activity Recommendations. http://www.w3.org/2002/ws/.

[W3C-HTML] HyperText Markup Language Home Page. http://www.w3.org/MarkUp, http://www.w3.org.

[Waller11] Waller, A., Sandy, I., Power, E., Aivaloglou, E., Skianis, C., Muñoz, A., Mana, A. "Policy Based Management for Security in Cloud Computing", STAVE 2011,1st International Workshop on Security & Trust for Applications in Virtualised Environments, J. Lopez, (Ed), June 2011, Loutraki, Greece, Springer CCIS.

[Wang10] Wang, M., Holub, V., Parsons, T., Murphy, J. and O'Sullivan, P. "Scalable run-time correlation engine for monitoring in a cloud computing environment". In Proceedings of the 2010 17th IEEE International Conference and Workshops on the Engineering of Computer-Based Systems, ECBS '10, p.p. 29–38, Washington, DC, USA, 2010. IEEE Computer Society.

[Wang04] Wang, X. et al. "Ontology-Based Context Modeling and Reasoning using OWL, Context". In Proceedings of Modeling and Reasoning Workshop at PerCom 2004.

[Ward97] Ward, A., Jones, A., Hopper, A. "A New Location Technique for the Active Office". IEEE Personal Communications Vol. 4, No. 5, p.p. 42–47, 1997.

[Wei03] Wei, Q., Farkas, K. Mendes, P., Phehofer, C., Nafisi, N. "Context-aware Handover Based on Active Network Technology" IWAN2003 Conference, Kyoto 10–12, December 2003.

[Weiser93] Weiser, Mark. "Ubiquitous Computing", IEEE Hot Topics, Vol. 26, p.p. 71–72, 1993.

[Wiederhold92] Wiederhold, G. "Mediators in the Architecture of Future Information Systems". In IEEE Computer Conference 1992.

[Winograd01] T. Winograd. "Architecture for Context". HCI Journal, 2001.

[Wolski99] Wolski, R., Spring, N. T., and Hayes, J. "The network weather service: a distributed resource performance forecasting service for metacomputing". Future Gener. Comput. Syst., 15:757–768, October 1999.

[Wong05] Wong, A., Ray, P., Parameswaran, N., Strassner, J., "Ontology mapping for the interoperability problem in network management", Journal on Selected Areas in Communications, Vol. 23, No. 10, p.p. 2058–2068, Oct. 2005.

X

[XML-RPC] XML-RPC, "XML-RPC specification", W3C Recommendation June 2003. http://www.xmlrpc.com/spec.

[XML-XSD] XML-XSD, "XML-XSD specification", W3C Recommendation, May 2001. http://www.w3.org/XML/Schema.

[XMLSPY] XML Schema Editor. http://www.xmlspy.com.

Y

[Yang03a] Yang, K., Galis, A., "Policy-driven mobile agents for context-aware service in next generation networks", IFIP 5th International Conference on Mobile Agents for Telecommunications (MATA 2003), Marrakech, ISBN 3-540-20298-6, Lecture Notes in Computer Science, Springer-Verlag, October 2003.

[Yang03b] Yang, K., Galis, A., "Network-centric context-aware service over integrated WLAN and GPRS networks", 14th IEE International Symposium On Personal, Indoor And Mobile Radio Communications, September 2003.

[Ying02] Ying, D., Mingshu, L. "TEMPPLET: A New Method for Domain-Specific Ontology Design" Lecture Notes in Computer Science-KNCS, Springer ISBN: 978-3-540-44222-6, January 2002.

Index